A Primer on
Memory Consistency
and Cache Coherence

Second Edition

Synthesis Lectures on Computer Architecture

Editors
Natalie Enright Jerger, *University of Toronto*
Margaret Martonosi, *Princeton University*

Founding Editor Emeritus
Mark D. Hill, *University of Wisconsin, Madison*

Synthesis Lectures on Computer Architecture publishes 50- to 100-page publications on topics pertaining to the science and art of designing, analyzing, selecting and interconnecting hardware components to create computers that meet functional, performance and cost goals. The scope will largely follow the purview of premier computer architecture conferences, such as ISCA, HPCA, MICRO, and ASPLOS.

A Primer on Memory Consistency and Cache Coherence, Second Edition

Vijay Nagarajan, Daniel J. Sorin, Mark D. Hill, and David A. Wood

ISBN: 978-3-031-01764-3 paperback
ISBN: 978-3-031-00061-4 ebook
ISBN: 978-3-031-02892-2 hardcover

DOI 10.1007/978-3-031-00061-4

A Publication in the Springer series
SYNTHESIS LECTURES ON ADVANCES IN AUTOMOTIVE TECHNOLOGY

Lecture #49
Series Editors: Natalie Enright Jerger, *University of Toronto*
                Margaret Martonosi, *Princeton University*
Founding Editor Emeritus: Mark D. Hill, *University of Wisconsin, Madison*
Series ISSN
Print 1935-3235   Electronic 1935-3243

A Primer on Memory Consistency and Cache Coherence

Second Edition

Vijay Nagarajan
University of Edinburgh

Daniel J. Sorin
Duke University

Mark D. Hill
University of Wisconsin, Madison

David A. Wood
University of Wisconsin, Madison

SYNTHESIS LECTURES ON COMPUTER ARCHITECTURE #49

ABSTRACT

Many modern computer systems, including homogeneous and heterogeneous architectures, support shared memory in hardware. In a shared memory system, each of the processor cores may read and write to a single shared address space. For a shared memory machine, the memory consistency model defines the architecturally visible behavior of its memory system. Consistency definitions provide rules about loads and stores (or memory reads and writes) and how they act upon memory. As part of supporting a memory consistency model, many machines also provide cache coherence protocols that ensure that multiple cached copies of data are kept up-to-date. The goal of this primer is to provide readers with a basic understanding of consistency and coherence. This understanding includes both the issues that must be solved as well as a variety of solutions. We present both high-level concepts as well as specific, concrete examples from real-world systems.

This second edition reflects a decade of advancements since the first edition and includes, among other more modest changes, two new chapters: one on consistency and coherence for non-CPU accelerators (with a focus on GPUs) and one that points to formal work and tools on consistency and coherence.

KEYWORDS

computer architecture, memory consistency, cache coherence, shared memory, memory systems, multicore processor, heterogeneous architecture, GPU, accelerators, semantics, verification

Contents

Preface to the Second Edition

This second edition of the Primer differs from the nearly decade-old first edition (2011) due to the addition of two chapters and modest other changes. New Chapter 10 addresses emerging work on non-CPU accelerators, principally general-purpose GPUs, that often implement consistency and coherence together. New Chapter 11 points to formal work and tools on consistency and coherence that have greatly advanced since the Primer's first edition. Other changes are more modest and include the following: Chapter 2 expands the definition of coherence to admit the GPU-like solutions of Chapter 10; Chapters 3 and 4 discuss recent advances in SC and TSO enforcement, respectively; and Chapter 5 adds a RISC-V case study.

The first edition of this work was supported in part by the U.S. National Science Foundation (CNS-0551401, CNS0720565, CCF-0916725, CCF-0444516, and CCF-0811290), Sandia/DOE (#MSN123960/ DOE890426), Semiconductor Research Corporation (contract 2009-HJ-1881), and the University of Wisconsin (Kellett Award to Hill). The views expressed herein are not necessarily those of the NSF, Sandia, DOE, or SRC. The second edition is supported by EPSRC under grant EP/M027317/1 to the University of Edinburgh, the U.S. National Science Foundation (CNS-1815656, CCF-1617824, CCF-1734706), and a John P. Morgridge Professorship to Hill. The views expressed herein are those of the authors only.

The authors thank Vasileios Gavrielatos, Olivier Giroux, Daniel Lustig, Kelly Shaw, Matthew Sinclair, and the students of the Spring 2019 edition of UW-Madison Graduate Parallel Computer Architecture (CS/ECE 757) taught by Hill for their insightful and detailed comments that improved this second edition, even as the authors take responsibility for its final content.

The authors also thank Margaret Martonosi, Natalie Enright Jerger, Michael Morgan, and their editorial team for enabling the production of this edition of the Primer.

Vijay thanks Nivi, Sattva, Swara, his parents, and in-laws for their love. Vijay also thanks IIT Madras for supporting his sabbatical, during which time most of the new chapters were written. In particular, he fondly reminisces the cricket nets sessions at the IIT Chemplast ground in the peak of Chennai "winter."

Dan thanks Deborah, Jason, and Julie for their love and for putting up with him taking the time to work on another edition of this Primer.

Mark wishes to thank Sue, Nicole, and Gregory for their love and support.

David thanks his coauthors for putting up with his deadline-challenged work style, his parents Roger and Ann Wood for inspiring him to be a second-generation Computer Sciences professor, and Jane, Alex, and Zach for helping him remember what life is all about.

Vijay Nagarajan, Daniel J. Sorin, Mark D. Hill, and David A. Wood
January 2020

Preface to the First Edition

This primer is intended for readers who have encountered memory consistency and cache coherence informally, but now want to understand what they entail in more detail. This audience includes computing industry professionals as well as junior graduate students.

We expect our readers to be familiar with the basics of computer architecture. Remembering the details of Tomasulo's algorithm or similar details is unnecessary, but we do expect readers to understand issues like architectural state, dynamic instruction scheduling (out-of-order execution), and how caches are used to reduce average latencies to access storage structures.

The primary goal of this primer is to provide readers with a basic understanding of consistency and coherence. This understanding includes both the issues that must be solved as well as a variety of solutions. We present both high-level concepts as well as specific, concrete examples from real-world systems. A secondary goal of this primer is to make readers aware of just how complicated consistency and coherence are. If readers simply discover what it is that they do not know—without actually learning it—that discovery is still a substantial benefit. Furthermore, because these topics are so vast and so complicated, it is beyond the scope of this primer to cover them exhaustively. It is not a goal of this primer to cover all topics in depth, but rather to cover the basics and apprise the readers of what topics they may wish to pursue in more depth.

We owe many thanks for the help and support we have received during the development of this primer. We thank Blake Hechtman for implementing and testing (and debugging!) all of the coherence protocols in this primer. As the reader will soon discover, coherence protocols are complicated, and we would not have trusted any protocol that we had not tested, so Blake's work was tremendously valuable. Blake implemented and tested all of these protocols using the Wisconsin GEMS simulation infrastructure [http://www.cs.wisc.edu/gems/].

For reviewing early drafts of this primer and for helpful discussions regarding various topics within the primer, we gratefully thank Trey Cain and Milo Martin. For providing additional feedback on the primer, we thank Newsha Ardalani, Arkaprava Basu, Brad Beckmann, Bob Cypher, Singh, Venkatanathan Varadarajan, Derek Williams, and Meng Zhang. While our reviewers provided great feedback, they may or may not agree with all of the final contents of this primer.

This work is supported in part by the National Science Foundation (CNS-0551401, CNS-0720565, CCF-0916725, CCF-0444516, and CCF-0811290), Sandia/DOE (#MSN123960/DOE890426), Semiconductor Research Corporation (contract 2009-HJ-1881), and the University of Wisconsin (Kellett Award to Hill). The views expressed herein are not necessarily those of the NSF, Sandia, DOE, or SRC.

Dan thanks Deborah, Jason, and Julie for their love and for putting up with him taking the time to work on another Synthesis Lecture. Dan thanks his Uncle Sol for helping inspire him to be an engineer in the first place. Lastly, Dan dedicates this book to the memory of Rusty Sneiderman, a treasured friend of thirty years who will be dearly missed by everyone who was lucky enough to have known him.

Mark wishes to thank Sue, Nicole, and Gregory for their love and support.

David thanks his coauthors for putting up with his deadline-challenged work style, his parents Roger and Ann Wood for inspiring him to be a second-generation Computer Sciences professor, and Jane, Alex, and Zach for helping him remember what life is all about.

Daniel J. Sorin, Mark D. Hill, and David A. Wood
November 2011

CHAPTER 1

Introduction to Consistency and Coherence

Many modern computer systems and most multicore chips (chip multiprocessors) support shared memory in hardware. In a shared memory system, each of the processor cores may read and write to a single shared address space. These designs seek various goodness properties, such as high performance, low power, and low cost. Of course, it is not valuable to provide these goodness properties without first providing correctness. Correct shared memory seems intuitive at a hand-wave level, but, as this lecture will help show, there are subtle issues in even defining what it means for a shared memory system to be correct, as well as many subtle corner cases in designing a correct shared memory implementation. Moreover, these subtleties must be mastered in hardware implementations where bug fixes are expensive. Even academics should master these subtleties to make it more likely that their proposed designs will work.

Designing and evaluating a correct shared memory system requires an architect to understand *memory consistency* and *cache coherence*, the two topics of this primer. Memory consistency (consistency, memory consistency model, or memory model) is a precise, architecturally visible definition of shared memory correctness. Consistency definitions provide rules about loads and stores (or memory reads and writes) and how they act upon memory. Ideally, consistency definitions would be simple and easy to understand. However, defining what it means for shared memory to behave correctly is more subtle than defining the correct behavior of, for example, a single-threaded processor core. The correctness criterion for a single processor core partitions behavior between one correct result and many incorrect alternatives. This is because the processor's architecture mandates that the execution of a thread transforms a given input state into a single well-defined output state, even on an out-of-order core. Shared memory consistency models, however, concern the loads and stores of multiple threads and usually allow many correct executions while disallowing many (more) incorrect ones. The possibility of multiple correct executions is due to the ISA allowing multiple threads to execute concurrently, often with many possible legal interleavings of instructions from different threads. The multitude of correct executions complicates the erstwhile simple challenge of determining whether an execution is correct. Nevertheless, consistency must be mastered to implement shared memory and, in some cases, to write correct programs that use it.

The microarchitecture—the hardware design of the processor cores and the shared memory system—must enforce the desired consistency model. As part of this consistency model support, the hardware provides cache coherence (or coherence). In a shared-memory system with caches, the cached values can potentially become out-of-date (or incoherent) when one of the processors updates its cached value. Coherence seeks to make the caches of a shared-memory system as functionally invisible as the caches in a single-core system; it does so by propagating a processor's write to other processors' caches. It is worth stressing that unlike consistency which is an architectural specification that defines shared memory correctness, coherence is a means to supporting a consistency model.

Even though consistency is the first major topic of this primer, we begin in Chapter 2 with a brief introduction to coherence because coherence protocols play an important role in providing consistency. The goal of Chapter 2 is to explain enough about coherence to understand how consistency models interact with coherent caches, but not to explore specific coherence protocols or implementations, which are topics we defer until the second portion of this primer in Chapters 6–9.

1.1   CONSISTENCY (A.K.A., MEMORY CONSISTENCY, MEMORY CONSISTENCY MODEL, OR MEMORY MODEL)

Consistency models define correct shared memory behavior in terms of loads and stores (memory reads and writes), without reference to caches or coherence. To gain some real-world intuition on why we need consistency models, consider a university that posts its course schedule online. Assume that the Computer Architecture course is originally scheduled to be in Room 152. The day before classes begin, the university registrar decides to move the class to Room 252. The registrar sends an e-mail message asking the website administrator to update the online schedule, and a few minutes later, the registrar sends a text message to all registered students to check the newly updated schedule. It is not hard to imagine a scenario—if, say, the website administrator is too busy to post the update immediately—in which a diligent student receives the text message, immediately checks the online schedule, and still observes the (old) class location Room 152. Even though the online schedule is eventually updated to Room 252 and the registrar performed the "writes" in the correct order, this diligent student observed them in a different order and thus went to the wrong room. A consistency model defines whether this behavior is correct (and thus whether a user must take other action to achieve the desired outcome) or incorrect (in which case the system must preclude these reorderings).

Although this contrived example used multiple media, similar behavior can happen in shared memory hardware with out-of-order processor cores, write buffers, prefetching, and multiple cache banks. Thus, we need to define shared memory correctness—that is, which shared memory behaviors are allowed—so that programmers know what to expect and implementors know the limits to what they can provide.

Shared memory correctness is specified by a memory consistency model or, more simply, a memory model. The memory model specifies the allowed behavior of multithreaded programs executing with shared memory. For a multithreaded program executing with specific input data, the memory model specifies what values dynamic loads may return and, optionally, what possible final states of the memory are. Unlike single-threaded execution, multiple correct behaviors are usually allowed, making understanding memory consistency models subtle.

Chapter 3 introduces the concept of memory consistency models and presents sequential consistency (SC), the strongest and most intuitive consistency model. The chapter begins by motivating the need to specify shared memory behavior and precisely defines what a memory consistency model is. It next delves into the intuitive SC model, which states that a multi-threaded execution should look like an interleaving of the sequential executions of each constituent thread, as if the threads were time-multiplexed on a single-core processor. Beyond this intuition, the chapter formalizes SC and explores implementing SC with coherence in both simple and aggressive ways, culminating with a MIPS R10000 case study.

In Chapter 4, we move beyond SC and focus on the memory consistency model implemented by x86 and historical SPARC systems. This consistency model, called total store order (TSO), is motivated by the desire to use first-in–first-out write buffers to hold the results of committed stores before writing the results to the caches. This optimization violates SC, yet promises enough performance benefit to inspire architectures to define TSO, which permits this optimization. In this chapter, we show how to formalize TSO from our SC formalization, how TSO affects implementations, and how SC and TSO compare.

Finally, Chapter 5 introduces "relaxed" or "weak" memory consistency models. It motivates these models by showing that most memory orderings in strong models are unnecessary. If a thread updates ten data items and then a synchronization flag, programmers usually do not care if the data items are updated in order with respect to each other but only that all data items are updated before the flag is updated. Relaxed models seek to capture this increased ordering flexibility to get higher performance or a simpler implementation. After providing this motivation, the chapter develops an example relaxed consistency model, called XC, wherein programmers get order only when they ask for it with a FENCE instruction (e.g., a FENCE after the last data update but before the flag write). The chapter then extends the formalism of the previous two chapters to handle XC and discusses how to implement XC (with considerable reordering between the cores and the coherence protocol). The chapter then discusses a way in which many programmers can avoid thinking about relaxed models directly: if they add enough FENCEs to ensure their program is data-race free (DRF), then most relaxed models will appear SC. With "SC for DRF," programmers can get both the (relatively) simple correctness model of SC with the (relatively) higher performance of XC. For those who want to reason more deeply, the chapter concludes by distinguishing acquires from releases, discussing write atomicity and causality, pointing to commercial examples (including an IBM Power case study), and touching upon high-level language models (Java and C++).

Returning to the real-world consistency example of the class schedule, we can observe that the combination of an email system, a human web administrator, and a text-messaging system represents an extremely weak consistency model. To prevent the problem of a diligent student going to the wrong room, the university registrar needed to perform a FENCE operation after her email to ensure that the online schedule was updated before sending the text message.

1.2   COHERENCE (A.K.A., CACHE COHERENCE)

Unless care is taken, a coherence problem can arise if multiple actors (e.g., multiple cores) have access to multiple copies of a datum (e.g., in multiple caches) and at least one access is a write. Consider an example that is similar to the memory consistency example. A student checks the online schedule of courses, observes that the Computer Architecture course is being held in Room 152 (reads the datum), and copies this information into her calendar app in her mobile phone (caches the datum). Subsequently, the university registrar decides to move the class to Room 252, updates the online schedule (writes to the datum) and informs the students via a text message. The student's copy of the datum is now stale, and we have an incoherent situation. If she goes to Room 152, she will fail to find her class. Examples of incoherence from the world of computing, but not including computer architecture, include stale web caches and programmers using un-updated code repositories.

Access to stale data (incoherence) is prevented using a coherence protocol, which is a set of rules implemented by the distributed set of actors within a system. Coherence protocols come in many variants but follow a few themes, as developed in Chapters 6–9. Essentially, all of the variants make one processor's write visible to the other processors by propagating the write to all caches, i.e., keeping the calendar in sync with the online schedule. But protocols differ in *when* and *how* the syncing happens. There are two major classes of coherence protocols. In the first approach, the coherence protocol ensures that writes are propagated to the caches synchronously. When the online schedule is updated, the coherence protocol ensures that the student's calendar is updated as well. In the second approach, the coherence protocol propagates writes to the caches asynchronously, while still honoring the consistency model. The coherence protocol does not guarantee that when the online schedule is updated, the new value will have propagated to the student's calendar as well; however, the protocol does ensure that the new value is propagated before the text message reaches her mobile phone. This primer focuses on the first class of coherence protocols (Chapters 6–9) while Chapter 10 discusses the emerging second class.

Chapter 6 presents the big picture of cache coherence protocols and sets the stage for the subsequent chapters on specific coherence protocols. This chapter covers issues shared by most coherence protocols, including the distributed operations of cache controllers and memory controllers and the common MOESI coherence states: modified (M), owned (O), exclusive (E), shared (S), and invalid (I). Importantly, this chapter also presents our table-driven methodology for presenting protocols with both stable (e.g., MOESI) and transient coherence states. Transient states are required in real implementations because modern systems rarely permit atomic transitions from one stable state to another (e.g., a read miss in state Invalid will spend some time waiting for a data response before it can enter state Shared). Much of the real complexity in coherence protocols hides in the transient states, similar to how much of processor core complexity hides in micro-architectural states.

Chapter 7 covers snooping cache coherence protocols, which initially dominated the commercial market. At the hand-wave level, snooping protocols are simple. When a cache miss occurs, a core's cache controller arbitrates for a shared bus and broadcasts its request. The shared bus ensures that all controllers observe all requests in the same order and thus all controllers can coordinate their individual, distributed actions to ensure that they maintain a globally consistent state. Snooping gets complicated, however, because systems may use multiple buses and modern buses do not atomically handle requests. Modern buses have queues for arbitration and can send responses that are unicast, delayed by pipelining, or out-of-order. All of these features lead to more transient coherence states. Chapter 7 concludes with case studies of the Sun UltraEnterprise E10000 and the IBM Power5.

Chapter 8 delves into directory cache coherence protocols that offer the promise of scaling to more processor cores and other actors than snooping protocols that rely on broadcast. There is a joke that all problems in computer science can be solved with a level of indirection. Directory protocols support this joke: A cache miss requests a memory location from the next level cache (or memory) controller, which maintains a directory that tracks which caches hold which locations. Based on the directory entry for the requested memory location, the controller sends a response message to the requestor or forwards the request message to one or more actors currently caching the memory location. Each message typically has one destination (i.e., no broadcast or multicast), but transient coherence states abound as transitions from one stable coherence state to another stable one can generate a number of messages proportional to the number of actors in the system. This chapter starts with a basic MSI directory protocol and then refines it to handle the MOESI states E and O, distributed directories, less stalling of requests, approximate directory entry representations, and more. The chapter also explores the design of the directory itself, including directory caching techniques. The chapter concludes with case studies of the old SGI Origin 2000 and the newer AMD HyperTransport, HyperTransport Assist, and Intel QuickPath Interconnect (QPI).

Chapter 9 deals with some, but not all, of the advanced topics in coherence. For ease of explanation, the prior chapters on coherence intentionally restrict themselves to the simplest system models needed to explain the fundamental issues. Chapter 9 delves into more complicated system models and optimizations, with a focus on issues that are common to both snooping and directory protocols. Initial topics include dealing with instruction caches, multilevel caches, write-through caches, translation lookaside buffers (TLBs), coherent direct memory access (DMA), virtual caches, and hierarchical coherence protocols. Finally, the chapter delves into performance optimizations (e.g., targeting migratory sharing and false sharing) and a new protocol family called Token Coherence that subsumes directory and snooping coherence.

1.3   CONSISTENCY AND COHERENCE FOR HETEROGENEOUS SYSTEMS

Modern computer systems are predominantly heterogeneous. A mobile phone processor today not only contains a multicore CPU, it also has a GPU and other accelerators (e.g., neural network hardware). In the quest for programmability, such heterogeneous systems are starting to support shared memory. Chapter 10 deals with consistency and coherence for such heterogeneous processors.

The chapter starts by focusing on GPUs, arguably the most popular accelerators today. The chapter observes that GPUs originally chose not to support hardware cache coherence, since GPUs are designed for embarrassingly parallel graphics workloads that do not synchronize or share data all that much. However, the absence of hardware cache coherence leads to programmability and/or performance challenges when GPUs are used for general-purpose workloads with fine-grained synchronization and data sharing. The chapter discusses in detail some of the promising coherence alternatives that overcome these limitations—in particular, explaining why the candidate protocols enforce the consistency model directly rather than implementing coherence in a consistency-agnostic manner. The chapter concludes with a brief discussion on consistency and coherence across CPUs and the accelerators.

1.4   SPECIFYING AND VALIDATING MEMORY CONSISTENCY MODELS AND CACHE COHERENCE

Consistency models and coherence protocols are complex and subtle. Yet, this complexity must be managed to ensure that multicores are programmable and that their designs can be validated. To achieve these goals, it is critical that consistency models are specified formally. A formal specification would enable programmers to clearly and exhaustively (with tool support) understand what behaviors are permitted by the memory model and what behaviors are not. Second, a precise formal specification is mandatory for validating implementations.

Chapter 11 starts by discussing two methods for specifying systems—axiomatic and operational—focusing on how these methods can be applied for consistency models and coherence protocols. Then the chapter goes over techniques for validating implementations—including processor pipeline and coherence protocol implementations—against their specification. The chapter discusses both formal methods and informal testing.

1.5   A CONSISTENCY AND COHERENCE QUIZ

It can be easy to convince oneself that one's knowledge of consistency and coherence is sufficient and that reading this primer is not necessary. To test whether this is the case, we offer this pop quiz.

Question 1: In a system that maintains sequential consistency, a core must issue coherence requests in program order. True or false? (Answer is in Section 3.8)

Question 2: The memory consistency model specifies the legal orderings of coherence transactions. True or false? (Section 3.8)

Question 3: To perform an atomic read–modify–write instruction (e.g., test-and-set), a core must always communicate with the other cores. True or false? (Section 3.9)

Question 4: In a TSO system with multithreaded cores, threads may bypass values out of the write buffer, regardless of which thread wrote the value. True or false? (Section 4.4)

Question 5: A programmer who writes properly synchronized code relative to the high-level language's consistency model (e.g., Java) does not need to consider the architecture's memory consistency model. True or false? (Section 5.9)

Question 6: In an MSI snooping protocol, a cache block may only be in one of three coherence states. True or false? (Section 7.2)

Question 7: A snooping cache coherence protocol requires the cores to communicate on a bus. True or false? (Section 7.6)

Question 8: GPUs do not support hardware cache coherence. Therefore, they are unable to enforce a memory consistency model. True or False? (Section 10.1).

Even though the answers are provided later in this primer, we encourage readers to try to answer the questions before looking ahead at the answers.

1.6   WHAT THIS PRIMER DOES NOT DO

This lecture is intended to be a primer on coherence and consistency. We expect this material could be covered in a graduate class in about ten 75-minute classes (e.g., one lecture per Chapter 2 to Chapter 11).

For this purpose, there are many things the primer does *not* cover. Some of these include the following.

- Synchronization. Coherence makes caches invisible. Consistency can make shared memory look like a single memory module. Nevertheless, programmers will probably need locks, barriers, and other synchronization techniques to make their programs useful. Readers are referred to the Synthesis Lecture on Shared-Memory synchronization [2].

- Commercial Relaxed Consistency Models. This primer does not cover the subtleties of the ARM, PowerPC, and RISC-V memory models, but does describe which mechanisms they provide to enforce order.

- Parallel programming. This primer does not discuss parallel programming models, methodologies, or tools.

- Consistency in distributed systems. This primer restricts itself to consistency within a shared memory multicore, and does not cover consistency models and their enforcement for a general distributed system. Readers are referred to the Synthesis Lectures on Database Replication [1] and Quorum Systems [3].

1.7   REFERENCES

[1] B. Kemme, R. Jiménez-Peris, and M. Patiño-Martínez. *Database Replication*. Synthesis Lectures on Data Management. Morgan & Claypool Publishers, 2010. DOI: 10.1007/978-1-4614-8265-9_110. 8

[2] M. L. Scott. *Shared-Memory Synchronization*. Synthesis Lectures on Computer Architecture. Morgan & Claypool Publishers, 2013. DOI: 10.2200/s00499ed1v01y201304cac023. 7

[3] M. Vukolic. *Quorum Systems: With Applications to Storage and Consensus*. Synthesis Lectures on Distributed Computing Theory. Morgan & Claypool Publishers, 2012. DOI: 10.2200/s00402ed1v01y201202dct009. 8

CHAPTER 2

Coherence Basics

In this chapter, we introduce enough about cache coherence to understand how consistency models interact with caches. We start in Section 2.1 by presenting the system model that we consider throughout this primer. To simplify the exposition in this chapter and the following chapters, we select the simplest possible system model that is sufficient for illustrating the important issues; we defer until Chapter 9 issues related to more complicated system models. Section 2.2 explains the cache coherence problem that must be solved and how the possibility of incoherence arises. Section 2.3 precisely defines cache coherence.

2.1 BASELINE SYSTEM MODEL

In this primer, we consider systems with multiple processor cores that share memory. That is, all cores can perform loads and stores to all (physical) addresses. The baseline system model includes a single multicore processor chip and off-chip main memory, as illustrated in Figure 2.1. The multicore processor chip consists of multiple single-threaded cores, each of which has its own private data cache, and a last-level cache (LLC) that is shared by all cores. Throughout this primer, when we use the term "cache," we are referring to a core's private data cache and not the LLC. Each core's data cache is accessed with physical addresses and is write-back. The cores and the LLC communicate with each other over an interconnection network. The LLC, despite being on the processor chip, is logically a "memory-side cache" and thus does not introduce another level of coherence issues. The LLC is logically just in front of the memory and serves to reduce the average latency of memory accesses and increase the memory's effective bandwidth. The LLC also serves as an on-chip memory controller.

This baseline system model omits many features that are common but that are not required for purposes of most of this primer. These features include instruction caches, multiple-level caches, caches shared among multiple cores, virtually addressed caches, TLBs, and coherent direct memory access (DMA). The baseline system model also omits the possibility of multiple multicore chips. We will discuss all of these features later, but for now, they would add unnecessary complexity.

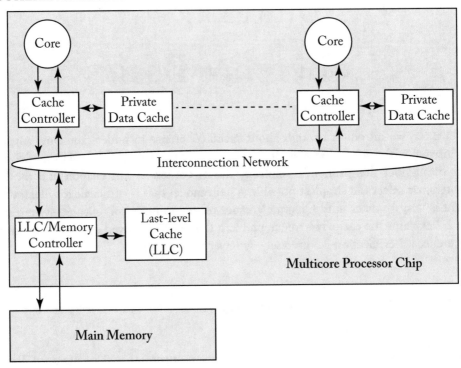

Figure 2.1: Baseline system model used throughout this primer.

2.2   THE PROBLEM: HOW INCOHERENCE COULD POSSIBLY OCCUR

The possibility of incoherence arises only because of one fundamental issue: there exist multiple actors with access to caches and memory. In modern systems, these actors are processor cores, DMA engines, and external devices that can read and/or write to caches and memory. In the rest of this primer, we generally focus on actors that are cores, but it is worth keeping in mind that other actors may exist.

Table 2.1 illustrates a simple example of incoherence. Initially, memory location A has the value 42 in memory as well as both of the cores' local caches. At time 1, Core 1 changes the value at memory location A from 42 to 43 in its cache, making Core 2's value of A in its cache stale. Core 2 executes a while loop loading, repeatedly, the (stale) value of A from its local cache. Clearly, this is an example of incoherence as the store from Core 1 has not not been made visible to Core 2 and consequently C2 is stuck in the while loop.

To prevent incoherence, the system must implement a *cache coherence protocol* that makes the store from Core 1 visible to Core 2. The design and implementation of these cache coherence protocols are the main topics of Chapters 6–9.

Table 2.1: Example of incoherence. Assume the value of memory at memory location A is initially 42 and cached in the local caches of both cores.

Time	Core C1	Core C2
1	S1: A = 43;	L1: while (A == 42);
2		L2: while (A == 42);
3		L3: while (A == 42);
4		…
n		Ln: while (A == 42);

2.3   THE CACHE COHERENCE INTERFACE

Informally, a coherence protocol must ensure that writes are made visible to all processors. In this section, we will more formally understand coherence protocols through the abstract interfaces they expose.

The processor cores interact with the coherence protocol through a coherence interface (Figure 2.2) that provides two methods: (1) a *read-request* method that takes in a memory location as the parameter and returns a value; and (2) a *write-request* method that takes in a memory location and a value (to be written) as parameters and returns an acknowledgment.

There are many coherence protocols that have appeared in the literature and been employed in real processors. We classify these protocols into two categories based on the nature of their coherence interfaces—specifically, based on whether there is a clean separation of coherence from the consistency model or whether they are indivisible.

Consistency-agnostic coherence. In the first category, a write is made visible to all other cores before returning. Because writes are propagated synchronously, the first category presents an interface that is identical to that of an atomic memory system (with no caches). Thus, any subsystem that interacts with the coherence protocol—e.g., the processor core pipeline—can assume it is interacting with an atomic memory system with no caches present. From a consistency enforcement perspective, this coherence interface enables a nice separation of concerns. The cache coherence protocol abstracts away the caches completely and presents an illusion of atomic memory—it is as if the caches are removed and only the memory is contained within the coherence box (Figure 2.2)—while the processor core pipeline enforces the orderings mandated by the consistency model specification.

Consistency-directed coherence. In the second, more-recent category, writes are propagated asynchronously—a write can thus return before it has been made visible to all processors, thus allowing for stale values (in real time) to be observed. However, in order to correctly enforce consistency, coherence protocols in this class must ensure that the order in which writes are

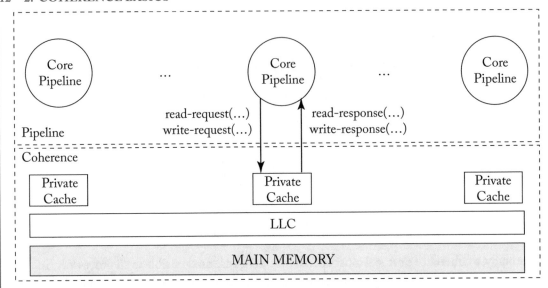

Figure 2.2: The pipeline-coherence interface.

eventually made visible adheres to the ordering rules mandated by the consistency model. Referring back to Figure 2.2, both the pipeline and the coherence protocol enforce the orderings mandated by the consistency model. This second category emerged to support throughput-based general-purpose graphics processing units (GP-GPUs) and gained prominence after the publication of the first edition of this primer.[1]

The primer (and the rest of the chapter) focuses on the first class of coherence protocols. We discuss the second class of coherence protocols in the context of heterogeneous coherence (Chapter 10).

2.4 (CONSISTENCY-AGNOSTIC) COHERENCE INVARIANTS

What invariants must a coherence protocol satisfy to make the caches invisible and present an abstraction of an atomic memory system?

There are several definitions of coherence that have appeared in textbooks and in published papers, and we do not wish to present all of them. Instead, we present the definition we prefer for its insight into the design of coherence protocols. In the sidebar, we discuss alternative definitions and how they relate to our preferred definition.

We define coherence through the *single-writer–multiple-reader (SWMR)* invariant. For any given memory location, at any given moment in time, there is either a single core that may

---

[1]For those of you concerned about the implications on consistency, note that it is possible to to enforce a variety of consistency models, including strong models such as SC and TSO, using this approach.

write it (and that may also read it) or some number of cores that may read it. Thus, there is never a time when a given memory location may be written by one core and simultaneously either read or written by any other cores. Another way to view this definition is to consider, for each memory location, that the memory location's lifetime is divided up into epochs. In each epoch, either a single core has read-write access or some number of cores (possibly zero) have read-only access. Figure 2.3 illustrates the lifetime of an example memory location, divided into four epochs that maintain the SWMR invariant.

Figure 2.3: Dividing a given memory location's lifetime into epochs.

In addition to the SWMR invariant, coherence requires that the value of a given memory location is propagated correctly. To explain why values matter, let us reconsider the example in Figure 2.3. Even though the SWMR invariant holds, if during the first read-only epoch Cores 2 and 5 can read different values, then the system is not coherent. Similarly, the system is incoherent if Core 1 fails to read the last value written by Core 3 during its read-write epoch or any of Cores 1, 2, or 3 fail to read the last write performed by Core 1 during its read-write epoch.

Thus, the definition of coherence must augment the SWMR invariant with a data value invariant that pertains to how values are propagated from one epoch to the next. This invariant states that the value of a memory location at the start of an epoch is the same as the value of the memory location at the end of its last read-write epoch.

There are other interpretations of these invariants that are equivalent. One notable example [5] interpreted the SMWR invariants in terms of tokens. The invariants are as follows. For each memory location, there exists a fixed number of tokens that is at least as large as the number of cores. If a core has all of the tokens, it may write the memory location. If a core has one or more tokens, it may read the memory location. At any given time, it is thus impossible for one core to be writing the memory location while any other core is reading or writing it.

Coherence invariants

1. **Single-Writer, Multiple-Read (SWMR) Invariant.** For any memory location A, at any given time, there exists only a single core that may write to A (and can also read it) or some number of cores that may only read A.

2. **Data-Value Invariant.** The value of the memory location at the start of an epoch is the same as the value of the memory location at the end of the its last read-write epoch.

2.4.1 MAINTAINING THE COHERENCE INVARIANTS

The coherence invariants presented in the previous section provide some intuition into how coherence protocols work. The vast majority of coherence protocols, called "invalidate protocols," are designed explicitly to maintain these invariants. If a core wants to read a memory location, it sends messages to the other cores to obtain the current value of the memory location and to ensure that no other cores have cached copies of the memory location in a read-write state. These messages end any active read-write epoch and begin a read-only epoch. If a core wants to write to a memory location, it sends messages to the other cores to obtain the current value of the memory location, if it does not already have a valid read-only cached copy, and to ensure that no other cores have cached copies of the memory location in either read-only or read-write states. These messages end any active read-write or read-only epoch and begin a new read-write epoch. This primer's chapters on cache coherence (Chapters 6–9) expand greatly upon this abstract description of invalidate protocols, but the basic intuition remains the same.

2.4.2 THE GRANULARITY OF COHERENCE

A core can perform loads and stores at various granularities, often ranging from 1–64 bytes. In theory, coherence could be performed at the finest load/store granularity. However, in practice, coherence is usually maintained at the granularity of cache blocks. That is, the hardware enforces coherence on a cache block by cache block basis. In practice, the SWMR invariant is likely to be that, for any *block* of memory, there is either a single writer or some number of readers. In typical systems, it is not possible for one core to be writing to the first byte of a block while another core is writing to another byte within that block. Although cache-block granularity is common, and it is what we assume throughout the rest of this primer, one should be aware that there have been protocols that have maintained coherence at finer and coarser granularities.

Sidebar: Consistency-Like Definitions of Coherence

Our preferred definition of coherence defines it from an implementation perspective—specifying hardware-enforced invariants regarding the access permissions of different cores to a memory location and the data values passed between cores.

There exists another class of definitions that defines coherence from a programmer's perspective, similar to how memory consistency models specify architecturally visible orderings of loads and stores.

One consistency-like approach to specifying coherence is related to the definition of sequential consistency. Sequential consistency (SC), a memory consistency model that we discuss in great depth in Chapter 3, specifies that the system must appear to execute all threads' loads and stores to all memory locations in a total order that respects the program order of each thread. Each load gets the value of the most recent store in that total order. A definition of coherence that is analogous to the definition of SC is that a coherent system must appear to execute all threads' loads and stores to a single memory location in a total order that respects the program order of each thread. This definition highlights an important distinction between coherence and consistency in the literature: coherence is specified on a per-memory location basis, whereas consistency is specified with respect to all memory locations. It is worth noting that any coherence protocol that satisfies the SWMR and data-value invariants (combined with a pipeline that does not reorder accesses to any specific location) is also guaranteed to satisfy this consistency-like definition of coherence. (However, the converse is not necessarily true.)

Another definition [1, 2] of coherence defines coherence with two invariants: (1) every store is eventually made visible to all cores and (2) writes to the same memory location are serialized (i.e., observed in the same order by all cores). IBM takes a similar view in the Power architecture [4], in part to facilitate implementations in which a sequence of stores by one core may have reached some cores (their values visible to loads by those cores) but not other cores. Invariant 2 is equivalent to the consistency-like definition we described earlier. In contrast to invariant 2, which is a *safety* invariant (bad things must not happen), invariant 1 is a *liveness* invariant (good things must eventually happen).

Another definition of coherence, as specified by Hennessy and Patterson [3], consists of three invariants: (1) a load to memory location A by a core obtains the value of the previous store to A by that core, unless another core has stored to A in between; (2) a load to A obtains the value of a store S to A by another core if S and the load "are sufficiently separated in time" and if no other store occurred between S and the load; and (3) stores to the same memory location are serialized (same as invariant 2 in the previous definition). Like the previous definition, this set of invariants captures both safety and liveness.

2.4.3   WHEN IS COHERENCE RELEVANT?

The definition of coherence—regardless of which definition we choose—is relevant only in certain situations, and architects must be aware of when it pertains and when it does not. We now discuss two important issues.

- Coherence pertains to *all* storage structures that hold blocks from the shared address space. These structures include the L1 data cache, L2 cache, shared last-level cache (LLC), and main memory. These structures also include the L1 instruction cache and translation lookaside buffers (TLBs).[2]

- Coherence is not *directly* visible to the programmer. Rather, the processor pipeline and coherence protocol jointly enforce the consistency model—and it is only the consistency model that is visible to the programmer.

---

[2]In some architectures, the TLB can hold mappings that are not strictly copies of blocks in shared memory.

2.5   REFERENCES

[1] K. Gharachorloo. Memory consistency models for shared-memory multiprocessors. Ph.D. thesis, Computer System Laboratory, Stanford University, December 1995. 15

[2] K. Gharachorloo, D. Lenoski, J. Laudon, P. Gibbons, A. Gupta, and J. Hennessy. Memory consistency and event ordering in scalable shared-memory. In *Proc. of the 17th Annual International Symposium on Computer Architecture*, pp. 15–26, May 1990. DOI: 10.1109/isca.1990.134503. 15

[3] J. L. Hennessy and D. A. Patterson. *Computer Architecture: A Quantitative Approach*, 4th ed. Morgan Kaufmann, 2007. 15

[4] IBM. Power ISA Version 2.06 Revision B. `http://www.power.org/resources/downloads/PowerISA_V2.06B_V2_PUBLIC.pdf`, July 2010. 15

[5] M. M. K. Martin, M. D. Hill, and D. A. Wood. Token coherence: Decoupling performance and correctness. In *Proc. of the 30th Annual International Symposium on Computer Architecture*, June 2003. DOI: 10.1109/isca.2003.1206999. 13

CHAPTER 3

Memory Consistency Motivation and Sequential Consistency

This chapter delves into memory consistency models (a.k.a. memory models) that define the behavior of shared memory systems for programmers and implementors. These models define correctness so that programmers know what to expect and implementors know what to provide. We first motivate the need to define memory behavior (Section 3.1), say what a memory consistency model should do (Section 3.2), and compare and contrast consistency and coherence (Section 3.3).

We then explore the (relatively) intuitive model of sequential consistency (SC). SC is important because it is what many programmers expect of shared memory and provides a foundation for understanding the more relaxed (weak) memory consistency models presented in the next two chapters. We first present the basic idea of SC (Section 3.4) and present a formalism of it that we will also use in subsequent chapters (Section 3.5). We then discuss implementations of SC, starting with naive implementations that serve as operational models (Section 3.6), a basic implementation of SC with cache coherence (Section 3.7), more optimized implementations of SC with cache coherence (Section 3.8), and the implementation of atomic operations (Section 3.9). We conclude our discussion of SC by providing a MIPS R10000 case study (Section 3.10) and pointing to some further reading (Section 3.11).

3.1 PROBLEMS WITH SHARED MEMORY BEHAVIOR

To see why shared memory behavior must be defined, consider the example execution of two cores[1] depicted in Table 3.1. (This example, as is the case for all examples in this chapter, assumes that the initial values of all variables are zero.) Most programmers would expect that core C2's register r2 should get the value NEW. Nevertheless, r2 can be 0 in some of today's computer systems.

Hardware can make r2 get the value 0 by reordering core C1's stores S1 and S2. Locally (i.e., if we look only at C1's execution and do not consider interactions with other threads), this reordering seems correct because S1 and S2 access different addresses. The sidebar on page 18

---

[1]Let "core" refer to software's view of a core, which may be an actual core or a thread context of a multithreaded core.

Table 3.1: Should r2 always be set to NEW?

Core C1	Core C2	Comments
S1: Store data = NEW; S2: Store flag = SET;	L1: Load r1 = flag; B1: if (r1 ≠ SET) goto L1; L2: Load r2 = data;	/* Initially, data = 0 & flag ≠ SET */ /* L1 & B1 may repeat many times */

Table 3.2: One possible execution of program in Table 3.1

Cycle	Core C1	Core C2	Coherence State of Data	Coherence State of Flag
1	S2: Store flag = SET		Read-only for C2	Read-write for C1
2		L1: Load r1=flag	Read-only for C2	Read-only for C2
3		L2: Load r2=data	Read-only for C2	Read-only for C2
4	S1: Store data = NEW		Read-write for C1	Read-only for C2

describes a few of the ways in which hardware might reorder memory accesses, including these stores. Readers who are not hardware experts may wish to trust that such reordering can happen (e.g., with a write buffer that is not first-in–first-out).

With the reordering of S1 and S2, the execution order may be S2, L1, L2, S1, as illustrated in Table 3.2.

Sidebar: How a Core Might Reorder Memory Access

This sidebar describes a few of the ways in which modern cores may reorder memory accesses to different addresses. Those unfamiliar with these hardware concepts may wish to skip this on first reading. Modern cores may reorder many memory accesses, but it suffices to reason about reordering two memory operations. In most cases, we need to reason only about a core reordering two memory operations to two different addresses, as the sequential execution (i.e., von Neumann) model generally requires that operations to the same address execute in the original program order. We break the possible reorderings down into three cases based on whether the reordered memory operations are loads or stores.

Store-store reordering. Two stores may be reordered if a core has a non-FIFO write buffer that lets stores depart in a different order than the order in which they entered. This might occur if the first store misses in the cache while the second hits or if the second store can coalesce with an earlier store (i.e., before the first store). Note that these reorderings are possible even if the core executes all instructions in program order. Reordering stores to different memory addresses has no effect on a single-threaded execution. However, in the multithreaded example of Table 3.1, reordering Core C1's stores allows Core C2 to see flag as SET before it sees the store to data. Note that the problem is not fixed even if the write buffer drains into a perfectly coherent memory hierarchy. Coherence will make all caches invisible, but the stores are already reordered.

Load-load reordering. Modern dynamically scheduled cores may execute instructions out of program order. In the example of Table 3.1, Core C2 could execute loads L1 and L2 out of order. Considering only a single-threaded execution, this reordering seems safe because L1 and L2 are to different addresses. However, reordering Core C2's loads behaves the same as reordering Core C1's stores; if the memory references execute in the order L2, S1, S2, and L1, then r2 is assigned 0. This scenario is even more plausible if the branch statement B1 is elided, so no control dependence separates L1 and L2.

Load-store and store-load reordering. Out-of-order cores may also reorder loads and stores (to different addresses) from the same thread. Reordering an earlier load with a later store (a load-store reordering) can cause many incorrect behaviors, such as loading a value after releasing the lock that protects it (if the store is the unlock operation). The example in Table 3.3 illustrates the effect of reordering an earlier store with a later load (a store-load reordering). Reordering Core C1's accesses S1 and L1 and Core C2's accesses S2 and L2 allows the counterintuitive result that both r1 and r2 are 0. Note that store-load reorderings may also arise due to local bypassing in the commonly implemented FIFO write buffer, even with a core that executes all instructions in program order.

A reader might assume that hardware should not permit some or all of these behaviors, but without a better understanding of what behaviors are allowed, it is hard to determine a list of what hardware can and cannot do.

Table 3.3: Can both r1 and r2 be set to 0?

Core C1	Core C2	Comments
S1: x = NEW;	S2: y = NEW;	/* Initially, x = 0 & y = 0 */
L1: r1 = y;	L2: r2 = x;	

This execution satisfies coherence because the SWMR property is not violated, so incoherence is not the underlying cause of this seemingly erroneous execution result.

Let us consider another important example inspired by Dekker's Algorithm for ensuring mutual exclusion, as depicted in Table 3.3. After execution, what values are allowed in r1 and r2? Intuitively, one might expect that there are three possibilities:

- (r1, r2) = (0, NEW) for execution S1, L1, S2, then L2

- (r1, r2) = (NEW, 0) for S2, L2, S1, and L1

- (r1, r2) = (NEW, NEW), e.g., for S1, S2, L1, and L2

Surprisingly, most real hardware, e.g., x86 systems from Intel and AMD, also allows (r1, r2) = (0, 0) because it uses first-in–first-out (FIFO) write buffers to enhance performance. As with the example in Table 3.1, all of these executions satisfy cache coherence, even (r1, r2) = (0, 0).

Some readers might object to this example because it is non-deterministic (multiple outcomes are allowed) and may be a confusing programming idiom. However, in the first place, all current multiprocessors are non-deterministic by default; all architectures of which we are aware permit multiple possible interleavings of the executions of concurrent threads. The illusion of determinism is sometimes, but not always, created by software with appropriate synchronization idioms. Thus, we must consider non-determinism when defining shared memory behavior.

Furthermore, memory behavior is usually defined for *all* executions of all programs, even those that are incorrect or intentionally subtle (e.g., for non-blocking synchronization algorithms). In Chapter 5, however, we will see some high-level language models that allow *some* executions to have undefined behavior, e.g., executions of programs with data races.

3.2   WHAT IS A MEMORY CONSISTENCY MODEL?

The examples in the previous sub-section illustrate that shared memory behavior is subtle, giving value to precisely defining (a) what behaviors programmers can expect and (b) what optimizations system implementors may use. A memory consistency model disambiguates these issues.

A *memory consistency model*, or, more simply, a *memory model*, is a specification of the allowed behavior of multithreaded programs executing with shared memory. For a multithreaded program executing with specific input data, it specifies what values dynamic loads may return. Unlike a single-threaded execution, multiple correct behaviors are usually allowed.

In general, a memory consistency model MC gives rules that partition executions into those obeying MC (*MC executions*) and those disobeying MC (*non-MC executions*). This partitioning of executions, in turn, partitions implementations. An *MC implementation* is a system that permits only MC executions, while a *non-MC implementation* sometimes permits non-MC executions.

Finally, we have been vague regarding the level of programming. We begin by assuming that programs are executables in a hardware instruction set architecture, and we assume that memory accesses are to memory locations identified by physical addresses (i.e., we are not considering the impact of virtual memory and address translation). In Chapter 5, we will discuss issues with high-level languages (HLLs). We will see then, for example, that a compiler allocating a variable to a register can affect an HLL memory model in a manner similar to hardware reordering memory references.

3.3 CONSISTENCY VS. COHERENCE

Chapter 2 defined cache coherence with two invariants that we informally repeat here. The SWMR invariant ensures that at any time for a memory location with a given address, either (a) one core may write (and read) the address or (b) zero or more cores may only read it. The *Data-Value Invariant* ensures that updates to the memory location are passed correctly so that cached copies of the memory location always contain the most recent version.

It may seem that cache coherence defines shared memory behavior. It does not. As we can see from Figure 3.1, the coherence protocol simply provides the processor core pipeline an abstraction of a memory system. It alone cannot determine shared memory behavior; the

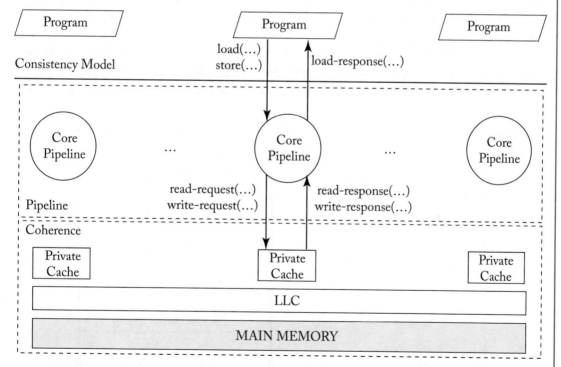

Figure 3.1: A consistency model is enforced by the processor core pipeline combined with the coherence protocol.

pipeline matters, too. If, for example, the pipeline reorders and presents memory operations to the coherence protocol in an order contrary to program order—even if the coherence protocol does its job correctly—shared memory correctness may not ensue.

In summary:

- Cache coherence does not equal memory consistency.

- A memory consistency implementation can use cache coherence as a useful "black box."

3.4 BASIC IDEA OF SEQUENTIAL CONSISTENCY (SC)

Arguably the most intuitive memory consistency model is SC. It was first formalized by Lamport [12], who called a single processor (core) *sequential* if "the result of an execution is the same as if the operations had been executed in the order specified by the program." He then called a multiprocessor *sequentially consistent* if "the result of any execution is the same as if the operations of all processors (cores) were executed in some sequential order, and the operations of each individual processor (core) appear in this sequence in the order specified by its program." This total order of operations is called *memory order*. In SC, memory order respects each core's program order, but other consistency models may permit memory orders that do not always respect the program orders.

Figure 3.2 depicts an execution of the example program from Table 3.1. The middle vertical downward arrow represents the memory order (<m) while each core's downward arrow represents its program order (<p). We denote memory order using the operator <m, so op1 <m

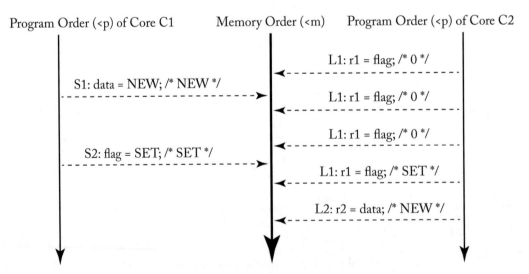

Figure 3.2: A sequentially consistent execution of Table 3.1's program.

op2 implies that op1 precedes op2 in memory order. Similarly, we use the operator <p to denote program order for a given core, so op1 <p op2 implies that op1 precedes op2 in that core's program order. Under SC, memory order *respects* each core's program order. "Respects" means that op1 <p op2 implies op1 <m op2. The values in comments (/* ... */) give the value loaded or stored. This execution terminates with r2 being NEW. More generally, all executions of Table 3.1's program terminate with r2 as NEW. The only non-determinism—how many times L1 loads flag as 0 before it loads the value SET once—is unimportant.

This example illustrates the value of SC. In Section 3.1, if you expected that r2 must be NEW, you were perhaps independently inventing SC, albeit less precisely than Lamport.

The value of SC is further revealed in Figure 3.3, which illustrates four executions of the program from Table 3.3. Figure 3.3a–c depict SC executions that correspond to the three intuitive outputs: (r1, r2) = (0, NEW), (NEW, 0), or (NEW, NEW). Note that Figure 3.3c depicts only one of the four possible SC executions that leads to (r1, r2) = (NEW, NEW); this execution is {S1, S2, L1, L2}, and the others are {S1, S2, L2, L1}, {S2, S1, L1, L2}, and {S2, S1, L2, L1}. Thus, across Figure 3.3a–c, there are six legal SC executions.

Figure 3.3d shows a non-SC execution corresponding to the output (r1, r2) = (0, 0). For this output, there is no way to create a memory order that respects program orders. Program order dictates that:

- S1 <p L1

- S2 <p L2

    But memory order dictates that:

- L1 <m S2 (so r1 is 0)

- L2 <m S1 (so r2 is 0)

Honoring all these constraints results in a cycle, which is inconsistent with a total order. The extra arcs in Figure 3.3d illustrate the cycle.

We have just seen six SC executions and one non-SC execution. This can help us understand SC implementations: an SC implementation must allow one or more of the first six executions, but cannot allow the seventh execution.

3.5   A LITTLE SC FORMALISM

In this section, we define SC more precisely, especially to allow us to compare SC with the weaker consistency models in the next two chapters. We adopt the formalism of Weaver and Germond [20]—an axiomatic method to specify consistency which we discuss more in Chapter 11—with the following notation: L(a) and S(a) represent a load and a store, respectively, to address a. Orders <p and <m define program and global memory order, respectively. Program

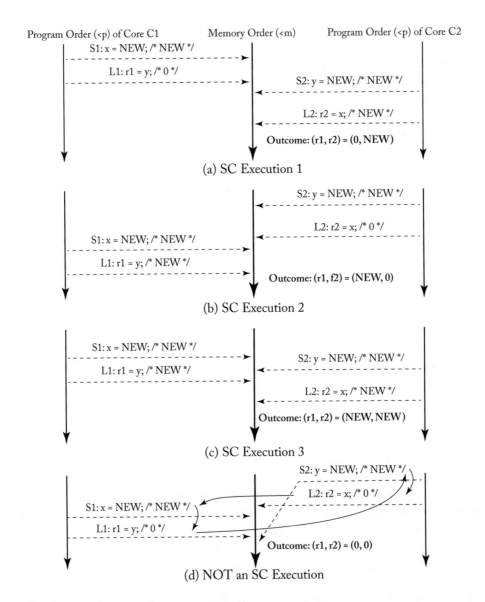

Figure 3.3: Four alternative executions of Table 3.3's program.

order <p is a per-core total order that captures the order in which each core logically (sequentially) executes memory operations. Global memory order <m is a total order on the memory operations of all cores.

An **SC execution** requires the following.

(1) All cores insert their loads and stores into the order <m respecting their program order, regardless of whether they are to the same or different addresses (i.e., a=b or $a \neq b$). There are four cases:

- If L(a) <p L(b) $\Rightarrow$ L(a) <m L(b) /* Load→Load */

- If L(a) <p S(b) $\Rightarrow$ L(a) <m S(b) /* Load→Store */

- If S(a) <p S(b) $\Rightarrow$ S(a) <m S(b) /* Store→Store */

- If S(a) <p L(b) $\Rightarrow$ S(a) <m L(b) /* Store→Load */

(2) Every load gets its value from the last store before it (in global memory order) to the same address:

Value of L(a) = Value of MAX $_{<m}$ {S(a) | S(a) <m L(a)}, where MAX $_{<m}$ denotes "latest in memory order."

Atomic read-modify-write (RMW) instructions, which we discuss in more depth in Section 3.9, further constrain allowed executions. Each execution of a test-and-set instruction, for example, requires that the load for the test and the store for the set logically appear consecutively in the memory order (i.e., no other memory operations for the same or different addresses interpose between them).

We summarize SC's ordering requirements in Table 3.4. The table specifies which program orderings are enforced by the consistency model. For example, if a given thread has a load before a store in program order (i.e., the load is "Operation 1" and the store is "Operation 2" in the table), then the table entry at this intersection is an "X" which denotes that these operations must be performed in program order. For SC, all memory operations must appear to perform in program order; under other consistency models, which we study in the next two chapters, some of these ordering constraints are relaxed (i.e., some entries in their ordering tables do not contain an "X").

An **SC implementation** permits only SC executions. Strictly speaking, this is the *safety* property for SC implementations (do no harm). SC implementations should also have some *liveness* properties (do some good). Specifically, a store must become eventually visible to a load that is repeatedly attempting to load that location. This property, referred to as eventual write-propagation, is typically ensured by the coherence protocol. More generally, starvation avoidance and some fairness are also valuable, but these issues are beyond the scope of this discussion.

Table 3.4: SC ordering rules. An "X" denotes an enforced ordering.

		Operation 2		
		Load	Store	RMW
Operation 1	Load	X	X	X
	Store	X	X	X
	RMW	X	X	X

3.6   NAIVE SC IMPLEMENTATIONS

SC permits two naive implementations that make it easier to understand which executions SC permits.

The Multitasking Uniprocessor

First, one can implement SC for multithreaded user-level software by executing all threads on a single sequential core (a uniprocessor). Thread T1's instructions execute on core C1 until a context switch to thread T2, etc. On a context switch, any pending memory operations must be completed before switching to the new thread. Because each thread's instructions in its quanta execute as one atomic block (and because the uniprocessor correctly honors memory dependencies), all of the SC rules are enforced.

The Switch

Second, one can implement SC with a set of cores C, a single switch, and memory, as depicted in Figure 3.4. Assume that each core presents memory operations to the switch one at a time in its program order. Each core can use any optimizations that do not affect the order in which it presents memory operations to the switch. For example, a simple five-stage in-order pipeline with branch prediction can be used.

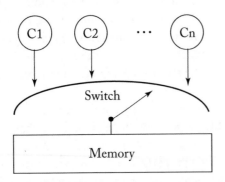

Each core Ci seeks to do its next memory access in its program order <p.

The switch selects one core, allows it to complete one memory access, and repeats; this defines memory order <m.

Figure 3.4: A simple SC implementation using a memory switch.

Assume next that the switch picks one core, allows memory to fully satisfy the load or store, and repeats this process as long as requests exist. The switch may pick cores by any method (e.g., random) that does not starve a core with a ready request. This implementation operationally implements SC by construction.

Assessment

The good news from these implementations is that they provide operational models defining (1) allowed SC executions and (2) SC implementation "gold standards." (In Chapter 11, we will see that such operational models can be used to formally specify consistency models.) The switch implementation also serves as an existence proof that SC can be implemented without caches or coherence.

The bad news, of course, is that the performance of these implementations does not scale up with increasing core count, due to the sequential bottleneck of using a single core in the first case and the single switch/memory in the second case. These bottlenecks have led some people to incorrectly conclude that SC precludes true parallel execution. It does not, as we will see next.

3.7   A BASIC SC IMPLEMENTATION WITH CACHE COHERENCE

Cache coherence facilitates SC implementations that can execute *non-conflicting* loads and stores—two operations *conflict* if they are to the same address and at least one of them is a store—completely in parallel. Moreover, creating such a system is conceptually simple.

Here, we treat coherence as mostly a black box that implements the SWMR invariant of Chapter 2. We provide some implementation intuition by opening the coherence block box slightly to reveal simple level-one (L1) caches that:

- use state *modified (M)* to denote an L1 block that one core can write and read,

- use state *shared (S)* to denote an L1 block that one or more cores can only read, and

- have *GetM* and *GetS* denote coherence requests to obtain a block in M and S, respectively.

We do not require a deep understanding of how coherence is implemented, as discussed in Chapter 6 and beyond.

Figure 3.5a depicts the model of Figure 3.4 with the switch and memory replaced by a cache-coherent memory system represented as a black box. Each core presents memory operations to the cache-coherent memory system one at a time in its program order. The memory system fully satisfies each request before beginning the next request for the same core.

Figure 3.5b "opens" the memory system black box a little to reveal that each core connects to its own L1 cache (we will talk about multithreading later). The memory system can respond to a load or store to block B if it has B with appropriate coherence permissions (state M or S for loads and M for stores). Moreover, the memory system can respond to requests from different

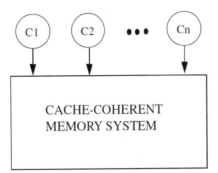

Each core Ci seeks to do its next memory access in its program order <p.

The memory system **logically** selects one core, allows it to complete one memory access, and repeats; this defines memory order <m.

(a) Black-box memory system

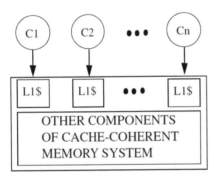

Same as above

Same as above, but cores can concurrently complete accesses to blocks with sufficient (L1) cache coherence permission, because such accesses must be non-conflicting (to different blocks or all loads) and may be placed into memory order in any logical order.

(b) Memory system with L1 caches exposed

Figure 3.5: Implementing SC with cache coherence.

cores in parallel, provided that the corresponding L1 caches have the appropriate permissions. For example, Figure 3.6a depicts the cache states before four cores each seek to do a memory operation. The four operations do not conflict, can be satisfied by their respective L1 caches, and therefore can be done concurrently. As depicted in Figure 3.6b, we can arbitrarily order these operations to obtain a legal SC execution model. More generally, operations that can be satisfied by L1 caches always can be done concurrently because coherence's SWMR invariant ensures they are non-conflicting.

Assessment

We have created an implementation of SC that:

- fully exploits the latency and bandwidth benefits of caches,

- is as scalable as the cache coherence protocol it uses, and

- decouples the complexities of implementing cores from implementing coherence.

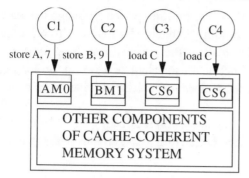

All four accesses can be executed
concurrently & be logically ordered.

Core C1's cache has block A in state
M (read-write) with value 0, C2 has
B in M with value 1, and both C3 and
C4 have C in S (read-only) with value 6.
(Of course real caches usually have
multi-word blocks.)

(a) Four accesses executed concurrently

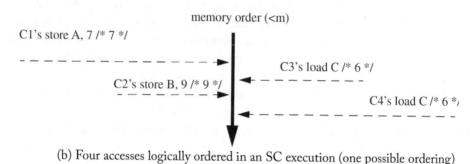

(b) Four accesses logically ordered in an SC execution (one possible ordering)

Figure 3.6: A concurrent SC execution with cache coherence.

3.8   OPTIMIZED SC IMPLEMENTATIONS WITH CACHE COHERENCE

Most real core implementations are more complicated than our basic SC implementation with
cache coherence. Cores employ features like prefetching, speculative execution, and multithreading in order to improve performance and tolerate memory access latencies. These features interact
with the memory interface, and we now discuss how these features impact the implementation
of SC. It is worth bearing in mind that any feature or optimization is legal as long as it does not
produce an end result (values returned by loads) that violates SC.

Non-Binding Prefetching

A non-binding prefetch for block B is a request to the coherent memory system to change B's
coherence state in one or more caches. Most commonly, prefetches are requested by software,
core hardware, or the cache hardware to change B's state in the level-one cache to permit loads
(e.g., B's state is M or S) or loads and stores (B's state is M) by issuing coherence requests such

as GetS and GetM. Importantly, in no case does a non-binding prefetch change the state of a register or data in block B. The effect of the non-binding prefetch is limited to within the "cache-coherent memory system" block of Figure 3.5a, making the effect of non-binding prefetches on the memory consistency model to be the functional equivalent of a no-op. So long as the loads and stores are performed in program order, it does not matter in what order coherence permissions are obtained.

Implementations may do non-binding prefetches without affecting the memory consistency model. This is useful for both internal cache prefetching (e.g., stream buffers) and more aggressive cores.

Speculative Cores

Consider a core that executes instructions in program order, but also does branch prediction wherein subsequent instructions, including loads and stores, begin execution, but may be squashed (i.e., have their effects nullified) on a branch misprediction. These squashed loads and stores can be made to look like non-binding prefetches, enabling this speculation to be correct because it has no effect on SC. A load after a branch prediction can be presented to the L1 cache, wherein it either misses (causing a non-binding GetS prefetch) or hits and then returns a value to a register. If the load is squashed, the core discards the register update, erasing any functional effect from the load—as if it never happened. The cache does not undo non-binding prefetches, as doing so is not necessary and prefetching the block can help performance if the load gets re-executed. For stores, the core may issue a non-binding GetM prefetch early, but it does not present the store to the cache until the store is guaranteed to commit.

Flashback to Quiz Question 1: In a system that maintains sequential consistency, a core must issue coherence requests in program order. *True or false?*
Answer: *False!* A core may issue coherence requests in any order.

Dynamically Scheduled Cores

Many modern cores dynamically schedule instruction execution out of program order to achieve greater performance than statically scheduled cores that must execute instructions in strict program order. A single-core processor that uses dynamic or out-of-(program-)order scheduling must simply enforce true data dependences within the program. However, in the context of a multicore processor, dynamic scheduling introduces a new issue: memory consistency speculation. Consider a core that wishes to dynamically reorder the execution of two loads, L1 and L2 (e.g., because L2's address is computed before L1's address). Many cores will speculatively execute L2 before L1, and they are predicting that this reordering is not visible to other cores, which would violate SC.

Speculating on SC requires that the core verify that the prediction is correct. Gharachorloo et al. [8] presented two techniques for performing this check. First, after the core speculatively executes L2, but before it commits L2, the core could check that the speculatively accessed block has not left the cache. So long as the block remains in the cache, its value could not have changed between the load's execution and its commit. To perform this check, the core tracks the address loaded by L2 and compares it to blocks evicted and to incoming coherence requests. An incoming GetM indicates that another core could observe L2 out of order, and this GetM would imply a mis-speculation and squash the speculative execution.

The second checking technique is to replay each speculative load when the core is ready to commit the load[2] [2, 17]. If the value loaded at commit does not equal the value that was previously loaded speculatively, then the prediction was incorrect. In the example, if the replayed load value of L2 is not the same as the originally loaded value of L2, then the load-load reordering has resulted in an observably different execution and the speculative execution must be squashed.

Non-Binding Prefetching in Dynamically Scheduled Cores

A dynamically scheduled core is likely to encounter load and store misses out of program order. For example, assume that program order is Load A, Store B, then Store C. The core may initiate non-binding prefetches "out of order," e.g., GetM C first and then GetS A and GetM B in parallel. SC is not affected by the order of non-binding prefetches. SC requires only that a core's loads and stores (appear to) access its level-one cache in program order. Coherence requires the level-one cache blocks to be in the appropriate states to receive loads and stores.

Importantly, SC (or any other memory consistency model):

- dictates the order in which loads and stores (appear to) get applied to coherent memory but

- does NOT dictate the order of coherence activity.

> **Flashback to Quiz Question 2:** The memory consistency model specifies the legal orderings of coherence transactions. *True or false?*
> **Answer:** *False!*

Multithreading

Multithreading—at coarse grain, fine grain, or simultaneous—can be accommodated by SC implementations. Each multithreaded core should be made logically equivalent to multiple (virtual) cores sharing each level-one cache via a switch where the cache chooses which virtual core to service next. Moreover, each cache can actually serve multiple non-conflicting requests concurrently because it can pretend that they were serviced in some order. One challenge is ensuring that a thread T1 cannot read a value written by another thread T2 on the same core before the store has been made "visible" to threads on other cores. Thus, while thread T1 may read the

---

[2]Roth [17] demonstrated a scheme for avoiding many load replays by determining when they are not necessary.

value as soon as thread T2 inserts the store in the memory order (e.g., by writing it to a cache block in state M), it cannot read the value from a shared load-store queue in the processor core.

Sidebar: Advanced SC Optimizations

This sidebar describes some advanced SC optimizations.

Post-retirement speculation. A single-core processor typically employs a structure called the write (store) buffer for hiding the latency of store misses; a store retires from the processor pipeline into the write buffer from where it drains into the cache/memory system off the critical path. This is safe on a single-core as long as loads check the write buffer for outstanding stores to the same address. On a multicore, however, the SC ordering rules preclude the naive use of a write buffer. Dynamically scheduled cores can hide some, but not all, of the store miss latency. To hide even more of the store miss latency, there have been many proposals for aggressive implementations of SC, utilizing speculation beyond the instruction window. The key idea is to speculatively retire loads and stores past pending store misses, while maintaining the state of speculatively retired instructions separately at either a fine granularity [9, 16] or coarse-grained chunks [1, 3, 11, 19].

Non-speculative reordering. It is even possible to non-speculatively perform memory operations out of order while enforcing SC, as long as the reordering is invisible to other cores [7, 18]. How do you ensure that reordering is invisible to other cores, without rollback recovery?

One approach (dubbed *coherence delaying*) involves delaying coherence requests: specifically, when a younger memory operation is retired past a pending older one, coherence requests to the younger one's location are delayed until the older memory operation retires. There is an inherent deadlock hazard with coherence delaying that necessitates careful deadlock avoidance mechanisms. In the example shown in Table 3.3, if both loads L1 and L2 retire past the stores and coherence requests to the respective locations are delayed, this can prevent either of the stores from completing, thereby leading to a deadlock.

Another approach (dubbed *predecessor serialization*) requires the older memory operation to do just enough—typically serializing at a central point—to ensure that it is safe for younger operations to complete past it. Conflict ordering [6] allows for loads and stores to retire past a pending store miss, as soon as the pending store serializes at the directory and determines a global list of pending stores; as long as the younger memory operation does not conflict with this list, it can safely retire. Gope and Lipasti [4] propose an approach tailored for in-order processors, wherein every load or store obtains a mutex from the directory in program order, but can retire out of order.

Finally, it is possible to leverage the help of the compiler or the memory management unit to determine accesses that can be safely reordered [5]. For example, two accesses to thread-private or read-only variables can be safely reordered.

3.9   ATOMIC OPERATIONS WITH SC

To write multithreaded code, a programmer needs to be able to synchronize the threads, and such synchronization often involves atomically performing pairs of operations. This functionality is provided by instructions that atomically perform a "read-modify-write" (RMW), such as the well-known "test-and-set," "fetch-and-increment," and "compare-and-swap." These atomic instructions are critical for proper synchronization and are used to implement spin-locks and other synchronization primitives. For a spin-lock, a programmer might use an RMW to atomically read whether the lock's value is unlocked (e.g., equal to 0) and write the locked value (e.g., equal to 1). For the RMW to be atomic, the read (load) and write (store) operations of the RMW must appear consecutively in the total order of operations required by SC.

Implementing atomic instructions in the microarchitecture is conceptually straightforward, but naive designs can lead to poor performance for atomic instructions. A correct but simplistic approach to implementing atomic instructions would be for the core to effectively lock the memory system (i.e., prevent other cores from issuing memory accesses) and perform its read, modify, and write operations to memory. This implementation, although correct and intuitive, sacrifices performance.

More aggressive implementations of RMWs leverage the insight that SC requires only the appearance of a total order of all requests. Thus, an atomic RMW can be implemented by first having a core obtain the block in state M in its cache, if the block is not already there in that state. The core then needs to only load and store the block in its cache—without any coherence messages or bus locking—as long as it waits to service any incoming coherence request for the block until after the store. This waiting does not risk deadlock because the store is guaranteed to complete.

Flashback to Quiz Question 3: To perform an atomic read-modify-write instruction (e.g., test-and-set), a core must always communicate with the other cores. *True or false?*
Answer: *False!*

An even more optimized implementation of RMWs could allow more time between when the load part and store part perform, without violating atomicity. Consider the case where the block is in a read-only state in the cache. The load part of the RMW can speculatively perform immediately, while the cache controller issues a coherence request to upgrade the block's state to read-write. When the block is then obtained in read-write state, the write part of the RMW performs. As long as the core can maintain the illusion of atomicity, this implementation is correct. To check whether the illusion of atomicity is maintained, the core must check whether the loaded block is evicted from the cache between the load part and the store part; this speculation support is the same as that needed for detecting mis-speculation in SC (Section 3.8).

3.10   PUTTING IT ALL TOGETHER: MIPS R10000

The MIPS R10000 [21] provides a venerable, but clean, commercial example for a speculative microprocessor that implements SC in cooperation with a cache-coherent memory hierarchy. Herein, we concentrate on aspects of the R10000 that pertain to implementing memory consistency.

The R10000 is a four-way superscalar RISC processor core with branch prediction and out-of-order execution. The chip supports writeback caches for L1 instructions and L1 data, as well as a private interface to an (off-chip) unified L2 cache.

The chip's main *system interface bus* supports cache coherence for up to four processors, as depicted in Figure 3.7 (adapted from Figure 1 in Yeager [21]). To construct an R10000-based system with more processors, such as the SGI Origin 2000 (discussed at length in Section 8.8.1), architects implemented a directory coherence protocol that connects R10000 processors via the system interface bus and a specialized Hub chip. In both cases, the R10000 processor core sees a coherent memory system that happens to be partially on-chip and partially off-chip.

During execution, an R10000 core issues (speculative) loads and stores in program order into an *address queue*. A load obtains a (speculative) value from the last store before it to the same address or, if none, the data cache. Loads and stores commit in program order and then remove their address queue entries. To commit a store, the L1 cache must hold the block in state M and the store's value must be written atomically with the commit.

Importantly, the eviction of a cache block—due to a coherence invalidation or to make room for another block—that contains a load's address in the address queue squashes the load and all subsequent instructions, which then re-execute. Thus, when a load finally commits, the loaded block was continuously in the cache between when it executed and when it commits, so it must get the same value as if it executed at commit. Because stores actually write to the cache at commit, the R10000 logically presents loads and stores in program order to the coherent memory system, thereby implementing SC, as discussed earlier.

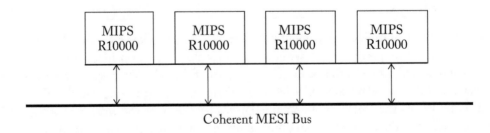

Figure 3.7: Coherent MESI bus connects up to four MIPS R10000 processors.

3.11 FURTHER READING REGARDING SC

Below we highlight a few of the papers from the vast literature surrounding SC.

Lamport [12] defined SC. As far as we know, Meixner and Sorin [15] were the first to prove that a system in which cores present loads and stores in program order to a cache coherent memory system was sufficient to implement SC, even as this result was intuitively believed for some time.

SC can be compared with database serializability [10]. The two concepts are similar in that they both insist that the operations from all entities appear to affect shared state in a serial order. The concepts differ due to the nature of and expectation for operations and shared state. With SC, each operation is a single memory access to volatile state (memory) that is assumed not to fail. With serializability, each operation is a transaction on a database that can read and write multiple database entities and is expected to obey ACID properties: Atomic—all or nothing even with failures; Consistent—leave the database consistent; Isolated—no effect from concurrent transactions; and Durable—effects survive crashes and power loss.

We followed Lamport and SPARC to define a total order of all memory accesses. While this can ease intuition for some, it is not necessary. Recall that two accesses *conflict* if they are from different threads, access the same location, and at least one is a store (or RMW). Instead of a total order, one can just define the constraints on conflicting accesses and leave non-conflicting accesses unordered, as pioneered by Shasha and Snir [18]. This view can be especially valuable for the relaxed models of Chapter 5.

Finally, a cautionary tale. We stated earlier (Section 3.7) that one way to check whether a speculatively executed load could have been observed out of order is to remember the value A that is speculatively read by a load and to commit the load if, at commit, the memory location has the same value A. Martin et al. [14] show that this is *not* the case for cores that perform value prediction [13]. With value prediction, when a load executes, the core can speculate on its value. Consider a core that speculates that a load of block X will produce the value A, although the value is actually B. Between when the core speculates on the load of X and when it replays the load at commit, another core changes block X's value to A. The core then replays the load at commit, compares the two values, which are equal, and mistakenly determines that the speculation was correct. The system can violate SC if it speculates in this way. This situation is analogous to the so-called ABA problem (http://en.wikipedia.org/wiki/ABA_problem), and Martin et al. showed that there are ways of checking speculation in the presence of value prediction that avoid the possibility of consistency violations (e.g., by also replaying all loads dependent on the initially speculated load). The point of this discussion is not to delve into the details of this particular corner case or its solutions, but rather to convince you to prove that your implementation is correct rather than rely on intuition.

3.12 REFERENCES

[1] C. Blundell, M. M. K. Martin, and T. F. Wenisch. InvisiFence: Performance-transparent memory ordering in conventional multiprocessors. In *Proc. of the 36th Annual International Symposium on Computer Architecture*, June 2009. DOI: 10.1145/1555754.1555785. 32

[2] H. W. Cain and M. H. Lipasti. Memory ordering: A value-based approach. In *Proc. of the 31st Annual International Symposium on Computer Architecture*, June 2004. DOI: 10.1109/isca.2004.1310766. 31

[3] L. Ceze, J. Tuck, P. Montesinos, and J. Torrellas. BulkSC: Bulk enforcement of sequential consistency. In *Proc. of the 34th Annual International Symposium on Computer Architecture*, June 2007. DOI: 10.1145/1250662.1250697. 32

[4] D. Gope and M. H. Lipasti. Atomic SC for simple in-order processors. In *20th IEEE International Symposium on High Performance Computer Architecture*, 2014. DOI: 10.1109/hpca.2014.6835950. 32

[5] A. Singh, S. Narayanasamy, D. Marino, T.D Millstein, and M. Musuvathi. End-to-end sequential consistency. In *39th International Symposium on Computer Architecture*, 2012. DOI: 10.1109/isca.2012.6237045. 32

[6] C. Lin, V. Nagarajan, R. Gupta, and B. Rajaram. Efficient sequential consistency via conflict ordering. In *Proc. of the 17th International Conference on Architectural Support for Programming Languages and Operating Systems ASPLOS*, 2012. DOI: 10.1145/2150976.2151006. 32

[7] K. Gharachorloo, S. V. Adve, A. Gupta, J. Hennessy, and M. D. Hill. Specifying system requirements for memory consistency models. *Technical Report CSL-TR93-594*, Stanford University, December 1993. 32

[8] K. Gharachorloo, A. Gupta, and J. Hennessy. Two techniques to enhance the performance of memory consistency models. In *Proc. of the International Conference on Parallel Processing*, vol. I, pp. 355–64, August 1991. 30

[9] C. Guiady, B. Falsafi, and T. Vijaykumar. Is SC + ILP = RC? In *Proc. of the 26th Annual International Symposium on Computer Architecture*, pp. 162–71, May 1999. DOI: 10.1109/isca.1999.765948. 32

[10] J. Gray and A. Reuter. *Transaction Processing: Concepts and Techniques*. Morgan Kaufmann Publishers, 1993. 35

[11] L. Hammond et al. Transactional memory coherence and consistency. In *Proc. of the 31st Annual International Symposium on Computer Architecture*, June 2004. DOI: 10.1109/isca.2004.1310767. 32

[12] L. Lamport. How to make a multiprocessor computer that correctly executes multiprocess programs. *IEEE Transactions on Computers*, C-28(9):690–91, September 1979. DOI: 10.1109/tc.1979.1675439. 22, 35

[13] M. H. Lipasti and J. P. Shen. Exceeding the dataflow limit via value prediction. In *Proc. of the 29th Annual IEEE/ACM International Symposium on Microarchitecture*, pp. 226–37, December 1996. DOI: 10.1109/micro.1996.566464. 35

[14] M. M. K. Martin, D. J. Sorin, H. W. Cain, M. D. Hill, and M. H. Lipasti. Correctly implementing value prediction in microprocessors that support multithreading or multiprocessing. In *Proc. of the 34th Annual IEEE/ACM International Symposium on Microarchitecture*, pp. 328–37, December 2001. DOI: 10.1109/micro.2001.991130. 35

[15] A. Meixner and D. J. Sorin. Dynamic verification of memory consistency in cache-coherent multithreaded computer architectures. In *Proc. of the International Conference on Dependable Systems and Networks*, pp. 73–82, June 2006. DOI: 10.1109/dsn.2006.29. 35

[16] P. Ranganathan, V. S. Pai, and S. V. Adve. Using speculative retirement and larger instruction windows to narrow the performance gap between memory consistency models. In *Proc. of the 9th ACM Symposium on Parallel Algorithms and Architectures*, pp. 199–210, June 1997. DOI: 10.1145/258492.258512. 32

[17] A. Roth. Store vulnerability window (SVW): Re-execution filtering for enhanced load optimization. In *Proc. of the 32nd Annual International Symposium on Computer Architecture*, June 2005. DOI: 10.1109/isca.2005.48. 31

[18] D. Shasha and M. Snir. Efficient and correct execution of parallel programs that share memory. *ACM Transactions on Programming Languages and Systems*, 10(2):282–312, April 1988. DOI: 10.1145/42190.42277. 32, 35

[19] T. F. Wenisch, A. Ailamaki, A. Moshovos, and B. Falsafi. Mechanisms for store-wait-free multiprocessors. In *Proc. of the 34th Annual International Symposium on Computer Architecture*, June 2007. DOI: 10.1145/1250662.1250696. 32

[20] D. L. Weaver and T. Germond, Eds. *SPARC Architecture Manual (Version 9)*. PTR Prentice Hall, 1994. 23

[21] K. C. Yeager. The MIPS R10000 superscalar microprocessor. *IEEE Micro*, 16(2):28–40, April 1996. DOI: 10.1109/40.491460. 34

CHAPTER 4

Total Store Order and the x86 Memory Model

A widely implemented memory consistency model is *total store order (TSO)*. TSO was first introduced by SPARC and, more importantly, appears to match the memory consistency model of the widely used x86 architecture. RISC-V also supports a TSO extension, RVTSO, in part to aid porting of code originally written for x86 or SPARC architectures. This chapter presents this important consistency model using a pattern similar to that in the previous chapter on sequential consistency. We first motivate TSO/x86 (Section 4.1) in part by pointing out limitations of SC. We then present TSO/x86 at an intuitive level (Section 4.2) before describing it more formally (Section 4.3), explaining how systems implement TSO/x86, including atomic instructions and instructions used to enforce ordering between instructions (Section 4.4). We conclude by discussing other resources for learning more about TSO/x86 (Section 4.5) and comparing TSO/x86 and SC (Section 4.6).

4.1 MOTIVATION FOR TSO/X86

Processor cores have long used *write (store) buffers* to hold committed (retired) stores until the rest of the memory system could process the stores. A store enters the write buffer when the store commits, and a store exits the write buffer when the block to be written is in the cache in a read-write coherence state. Significantly, a store can enter the write buffer before the cache has obtained read-write coherence permissions for the block to be written; the write buffer thus hides the latency of servicing a store miss. Because stores are common, being able to avoid stalling on most of them is an important benefit. Moreover, it seems sensible to not stall the core because the core does not need anything, as the store seeks to update memory but not core state.

For a single-core processor, a write buffer can be made architecturally invisible by ensuring that a load to address A returns the value of the most recent store to A even if one or more stores to A are in the write buffer. This is typically done by either *bypassing* the value of the most recent store to A to the load from A, where "most recent" is determined by program order, or by stalling a load of A if a store to A is in the write buffer.

When building a multicore processor, it seems natural to use multiple cores, each with its own bypassing write buffer, and assume that the write buffers continue to be architecturally invisible.

Table 4.1: Can both r1 and r2 be set to 0?

Core C1	Core C2	Comments
S1: x = NEW;	S2: y = NEW;	/* Initially, x = 0 & y = 0 */
L2: r1 = y;	L2: r2 = x;	

This assumption is wrong. Consider the example code in Table 4.1 (which is the same as Table 3.3 in the previous chapter). Assume a multicore processor with in-order cores, where each core has a single-entry write buffer and executes the code in the following sequence.

- Core C1 executes store S1, but buffers the newly stored NEW value in its write buffer.

- Likewise, core C2 executes store S2 and holds the newly stored NEW value in its write buffer.

- Next, both cores perform their respective loads, L1 and L2, and obtain the old values of 0.

- Finally, both cores' write buffers update memory with the newly stored values NEW.

The net result is that $(r1, r2) = (0, 0)$. As we saw in the previous chapter, this is an execution result forbidden by SC. Without write buffers, the hardware is SC, but with write buffers, it is not, making write buffers architecturally visible in a multicore processor.

One response to write buffers being visible would be to turn them off, but vendors have been loath to do this because of the potential performance impact. Another option is to use aggressive, speculative SC implementations that make write buffers invisible again, but doing so adds complexity and can waste power to both detect violations and handle mis-speculations.

The option chosen by SPARC and later x86 was to abandon SC in favor of a memory consistency model that allows straightforward use of a first-in–first-out (FIFO) write buffer at each core. The new model, TSO, allows the outcome "$(r1, r2) = (0, 0)$." This model astonishes some people but, it turns out, behaves like SC for most programming idioms and is well defined in all cases.

4.2   BASIC IDEA OF TSO/X86

As execution proceeds, SC requires that each core preserves the program order of its loads and stores for all four combinations of consecutive operations:

- Load → Load

- Load → Store

- Store → Store

- Store → Load /* Included for SC but omitted for TSO */

TSO includes the first three constraints but not the fourth. This omission does not matter for most programs. Table 4.2 repeats the example program of Table 3.1 in the previous chapter. In this case, TSO allows the same executions as SC because TSO preserves the order of core C1's two stores and core C2's two (or more) loads. Figure 4.1 (the same as Figure 3.2 in the previous chapter) illustrates the execution of this program.

Table 4.2: Should r2 always be set to NEW?

Core C1	Core C2	Comments
S1: Store data = NEW; S2: Store flag = SET;		/* Initially, data = 0 & flag ≠ SET */
	L1: Load r1 = flag B1: if (r1 ≠ SET) goto L1; L2: Load r2=data;	/* L1 & B1 may repeat many times */

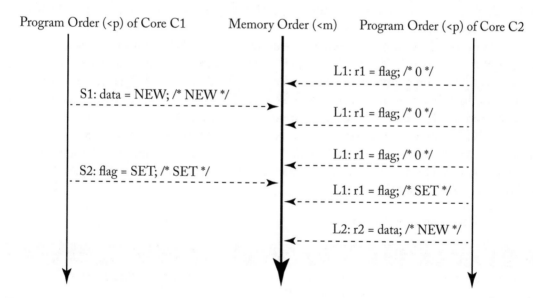

Figure 4.1: A TSO execution of Table 4.2's program.

More generally, TSO behaves the same as SC for common programming idioms that follow:

- C1 loads and stores to memory locations D1, . . ., Dn (often data),

- C1 stores to F (often a synchronization flag) to indicate that the above work is complete,

- C2 loads from F to observe the above work is complete (sometimes spinning first and often using a RMW instruction), and

- C2 loads and stores to some or all of the memory locations D1, . . ., Dn.

TSO, however, allows some non-SC executions. Under TSO, the program from Table 4.1 (repeat of Table 3.3 from the last chapter) allows all four outcomes depicted in Figure 4.2. Under SC, only the first three are legal outcomes (as depicted in Figure 3.3 of the last chapter). The execution in Figure 4.2d illustrates an execution that conforms to TSO but violates SC by not honoring the fourth (i.e., Store → Load) constraint. Omitting the fourth constraint allows each core to use a write buffer. Note that the third constraint means that the write buffer must be FIFO (and not, for example, coalescing) to preserve store–store order.

Programmers (or compilers) can prevent the execution in Figure 4.2d by inserting a FENCE instruction between S1 and L1 on core C1 and between S2 and L2 on core C2. Executing a FENCE on core Ci ensures that Ci's memory operations before the FENCE (in program order) get placed in memory order before Ci's memory operations after the FENCE. FENCEs (a.k.a. memory barriers) are rarely used by programmers using TSO because TSO "does the right thing" for most programs. Nevertheless, FENCEs play an important role for the relaxed models discussed in the next chapter.

TSO does allow some non-intuitive execution results. Table 4.3 illustrates a modified version of the program in Table 4.1 in which cores C1 and C2 make local copies of x and y, respectively. Many programmers might assume that if both r2 and r4 equal 0, then r1 and r3 should also be 0 because the stores S1 and S2 must be inserted into memory order after the loads L2 and L4. However, Figure 4.3 illustrates an execution that shows r1 and r3 bypassing the value NEW from the per-core write buffers. In fact, to preserve single-thread sequential semantics, each core must see the effect of its own store in program order, even though the store is not yet observed by other cores. Thus, under all TSO executions, the local copies r1 and r3 will always be set to the NEW value.

Table 4.3: Can r1 or r3 be set to 0?

Core C1	Core C2	Comments
S1: x = NEW;	S2: y = NEW;	/* Initially, x = 0 & y = 0 */
L1: r1 = x;	L3: r3 = y;	
L2: r2 = y;	L4: r4 = x;	/* Assume r2 = 0 & r4 =0 */

4.3   A LITTLE TSO/X86 FORMALISM

In this section we define TSO more precisely with a definition that makes only three changes to the SC definition of Section 3.5.

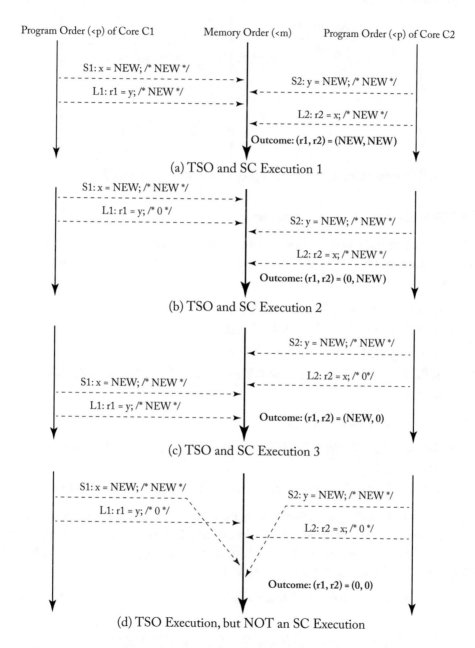

Figure 4.2: Four alternative TSO executions of Table 4.1's program.

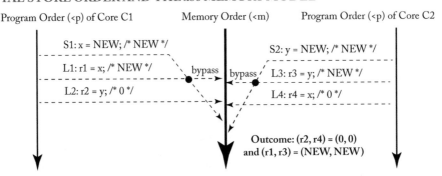

Figure 4.3: A TSO execution of Table 4.3's program (with "bypassing").

A **TSO execution** requires the following.

1. All cores insert their loads and stores into the memory order <m respecting their program order, regardless of whether they are to the same or different addresses (i.e., a==b or a!=b). There are four cases:

   - If L(a) <p L(b) ⇒ L(a) <m L(b) /* Load → Load */
   - If L(a) <p S(b) ⇒ L(a) <m S(b) /* Load → Store */
   - If S(a) <p S(b) ⇒ S(a) <m S(b) /* Store → Store */
   - ~~If S(a) <p L(b) ⇒ S(a) <m L(b) /* Store → Load */~~ /* **Change 1: Enable FIFO Write Buffer */**

2. Every load gets its value from the last store before it to the same address:

   - ~~Value of L(a) = Value of MAX <m {S(a) | S(a) <m L(a)}~~ /* **Change 2: Need Bypassing */**
   - Value of L(a) = Value of MAX <m {S(a) | S(a) <m L(a) **or** S(a) <p L(a)}

   This last mind-bending equation says that the value of a load is the value of the last store to the same address that is either (a) before it in memory order or (b) before it in program order (but possibly after it in memory order), with option (b) taking precedence (i.e., write buffer bypassing overrides the rest of the memory system).

3. Part (1) must be augmented to define FENCEs: /* **Change 3: FENCEs Order Everything** */

   - If L(a) <p FENCE ⇒ L(a) <m FENCE /* Load → FENCE */
   - If S(a) <p FENCE ⇒ S(a) <m FENCE /* Store → FENCE */
   - If FENCE <p FENCE ⇒ FENCE <m FENCE /* FENCE → FENCE */

- If FENCE <p L(a) ⇒ FENCE <m L(a) /* FENCE → Load */

- If FENCE <p S(a) ⇒ FENCE <m S(a) /* FENCE → Store */

Because TSO already requires all but the Store → Load order, one can alternatively define TSO FENCEs as only ordering:

- If S(a) <p FENCE ⇒ S(a) <m FENCE /* Store → FENCE */

- If FENCE <p L(a) ⇒ FENCE <m L(a) /* FENCE → Load */

We choose to have TSO FENCEs redundantly order everything because doing so does not hurt and makes them like the FENCEs we define for more relaxed models in the next chapter.

We summarize TSO's ordering rules in Table 4.4. This table has two important differences from the analogous table for SC (Table 3.4). First, if Operation #1 is a store and Operation #2 is a load, the entry at that intersection is a "B" instead of an "X"; if these operations are to the same address, the load must obtain the value just stored even if the operations enter memory order out of program order. Second, the table includes FENCEs, which were not necessary in SC; an SC system behaves as if there is already a FENCE before and after every operation.

Table 4.4: TSO ordering rules. An "X" denotes an enforced ordering. A "B" denotes that bypassing is required if the operations are to the same address. Entries that are different from the SC ordering rules are shaded and shown in bold.

		Operation 2			
		Load	Store	RMW	FENCE
Operation 1	Load	X	X	X	**X**
	Store	**B**	X	X	**X**
	RMW	X	X	X	**X**
	FENCE	**X**	**X**	**X**	X

It is widely believed that the x86 memory model is equivalent to TSO (for normal cacheable memory and normal instructions), but to the best of our knowledge, neither AMD nor Intel have guaranteed this or released a formal x86 memory model specification. AMD and Intel publicly define the x86 memory model with examples and prose in a process that is well summarized in Section 2 of Sewell et al. [15]. All examples conform to TSO, and all prose seems consistent with TSO. This equivalence can be proven only if a public, formal description of the x86 memory model were made available. This equivalence could be disproved if counter-example(s) showed an x86 execution not allowed by TSO, a TSO execution not allowed by x86, or both.

That x86 is equivalent to TSO is supported by Sewell et al. [15], summarized in CACM with more details elsewhere [9, 14]. In particular, the authors propose the x86-TSO model. The model has two forms that the authors prove equivalent. The first form provides an abstract machine that resembles Figure 4.4a of the next section with the addition of a single global lock for modeling x86 LOCK'd instructions. The second form is a labeled transition system. The first form makes the model accessible to practitioners while the latter eases formal proofs. On one hand, x86-TSO appears consistent with the informal rules and litmus tests in x86 specifications. On the other hand, empirical tests on several AMD and Intel platforms did not reveal any violations of the x86-TSO model (but this is no proof that they cannot). In summary, like Sewell et al., we urge creators of x86 hardware and software to adopt the unambiguous and accessible x86-TSO model.

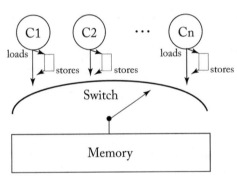

This implementation is the same as for Figure 3.4, except that each core Ci has a FIFO write buffer that buffers stores until they go to memory.

(a) A TSO implementation using a switch

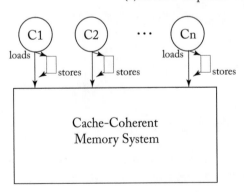

This TSO implementation replaces the switch above with a cache-coherent memory system in a manner analogous to what was done for SC.

(b) A TSO implementation using cache coherence

Figure 4.4: Two TSO implementations.

4.4   IMPLEMENTING TSO/X86

The implementation story for TSO/x86 is similar to SC with the addition of per-core FIFO write buffers. Figure 4.4a updates the switch of Figure 3.4 to accommodate TSO and operates as follows.

- Loads and stores leave each core in that core's program order <p.

- A load either bypasses a value from the write buffer or awaits the switch as before.

- A store enters the tail of the FIFO write buffer or stalls the core if the buffer is full.

- When the switch selects core Ci, it performs either the next load or the store at the head of the write buffer.

In Section 3.7, we showed that, for SC, the switch can be replaced by a cache coherent memory system and then argued that cores could be speculative and/or multithreaded and that nonbinding prefetches could be initiated by cores, caches, or software.

As illustrated in Figure 4.4b, the *same* argument holds for TSO with a FIFO writer buffer interposed between each core and the cache-coherent memory system. Thus, aside from the write buffer, all the previous SC implementation discussion holds for TSO and provides a way to build TSO implementations. Moreover, most current TSO implementations seem to use only the above approach: take an SC implementation and insert write buffers.

Regarding the write buffer, the literature and product space for how exactly speculative cores implement them is beyond the scope of this chapter. For example, microarchitectures can physically combine the store queue (uncommitted stores) and write buffer (committed stores), and/or physically separate load and store queues.

Finally, multithreading introduces a subtle write buffer issue for TSO. TSO write buffers are logically private to each thread context (virtual core). Thus, on a multithreaded core, one thread context should never bypass from the write buffer of another thread context. This logical separation can be implemented with per-thread-context write buffers or, more commonly, by using a shared write buffer with entries tagged by thread-context identifiers that permit bypassing only when tags match.

Flashback to Quiz Question 4: In a TSO system with multithreaded cores, threads may bypass values out of the write buffer, regardless of which thread wrote the value. *True or false?* **Answer:** *False!* A thread may bypass values that it has written, but other threads may not see the value until the store is inserted into the memory order.

4.4.1   IMPLEMENTING ATOMIC INSTRUCTIONS

The implementation issues for atomic RMW instructions in TSO are similar to those for atomic instructions for SC. The key difference is that TSO allows loads to pass (i.e., be ordered before) earlier stores that have been written to a write buffer. The impact on RMWs is that the "write" (i.e., store) may be written to the write buffer.

To understand the implementation of atomic RMWs in TSO, we consider the RMW as a load immediately followed by a store.

The load part of the RMW cannot pass earlier loads due to TSO's ordering rules. It might at first appear that the load part of the RMW could pass earlier stores in the write buffer, but this is not legal. If the load part of the RMW passes an earlier store, then the store part of the RMW would also have to pass the earlier store because the RMW is an atomic pair. But because stores are not allowed to pass each other in TSO, the load part of the RMW cannot pass an earlier store either.

These ordering constraints on RMWs impact the implementation. Because the load part of the RMW cannot be performed until earlier stores have been ordered (i.e., exited the write buffer), the atomic RMW effectively drains the write buffer before it can perform the load part of the RMW. Furthermore, to ensure that the store part can be ordered immediately after the load part, the load part requires read-write coherence permissions, not just the read permissions that suffice for normal loads. Lastly, to guarantee atomicity for the RMW, the cache controller may not relinquish coherence permission to the block between the load and the store.

More optimized implementations of RMWs are possible. For example, the write buffer does not need to be drained as long as (a) every entry already in the write buffer has read-write permission in the cache and maintains the read-write permission in the cache until the RMW commits, and (b) the core performs MIPS R10000-style checking of load speculation (Section 3.8). Logically, all of the earlier stores and loads would then commit as a unit (sometimes called a "chunk") immediately before the RMW.

4.4.2 IMPLEMENTING FENCES

Systems that support TSO do not provide ordering between a store and a subsequent (in program order) load, although they do require the load to get the value of the earlier store. In situations in which the programmer wants those instructions to be ordered, the programmer must explicitly specify that ordering by putting a FENCE instruction between the store and the subsequent load. The semantics of the FENCE specify that all instructions before the FENCE in program order must be ordered before any instructions after the FENCE in program order. For systems that support TSO, the FENCE thus prohibits a load from bypassing an earlier store. In Table 4.5, we revisit the example from Table 4.1, but we have added two FENCE instructions that were not present earlier. Without these FENCEs, the two loads (L1 and L2) can bypass the two stores (S1 and S2), leading to an execution in which r1 and r2 both get set to zero. The added FENCEs prohibit that reordering and thus prohibit that execution.

Because TSO permits only one type of reordering, FENCEs are fairly infrequent and the implementation of FENCE instructions is not too critical. A simple implementation—such as

Table 4.5: Can both r1 and r2 be set to 0?

Core C1	Core C2	Comments
S1: x = NEW;	S2: y = NEW;	/* Initially, x = 0 & y = 0 */
FENCE	**FENCE**	
L1: r1 = y;	L2: r2 = x;	

draining the write buffer when a FENCE is executed and not permitting subsequent loads to execute until an earlier FENCE has committed—may provide acceptable performance.

However, for consistency models that permit far more reordering (discussed in the next chapter), FENCE instructions are more frequent and their implementations can have a significant impact on performance.

Sidebar: Non-speculative TSO Optimizations

This sidebar describes some advanced non-speculative TSO optimizations.

Non-speculative TSO reordering. There have been papers that have shown that both loads [11, 12] and stores [13] can be reordered non-speculatively, while still enforcing TSO, using coherence delaying. As mentioned earlier, the key challenge is to ensure that the delays do not cause cyclic dependencies that lead to a deadlock, which all of the above papers address.

RMW without write buffer drain. In Section 4.4.1, we showed how to move the write buffer drain off the critical path using speculation. Rajaram et al. [10] show that the same effect can be achieved non-speculatively if the atomicity semantics of an RMW are redefined. Recall that for the RMW to be atomic, we earlier mandated that the read and the write operations of the RMW must appear consecutively in the TSO global memory order. Consider the following relaxation in which an RMW is deemed to be atomic as long as writes to the same address as that of the RMW do not appear between the read and write in the global memory order. Note that this relaxed definition matches the intuitive definition of an RMW and is sufficient for them to be used in synchronization situations. At the same time, it allows for an RMW implementation in which the load part of the RMW can bypass the earlier stores in the write buffer without requiring MIPS R10000-style load speculation checks. However, ensuring RMW atomicity necessitates coherence delaying until the write buffer drains, which introduces a deadlock hazard. Rajaram et al. [10] show how deadlocks can be avoided.

Reordering past a FENCE. There have been several papers that have proposed optimized FENCE implementations [3, 4, 6, 8] that enable memory operations following the FENCE to be non-speculatively retired before those that precede the FENCE. These techniques either use coherence delays or predecessor serialization or a combination of both.

4.5   FURTHER READING REGARDING TSO

Collier [2] characterized alternative memory consistency models, including that of the IBM System/370, via a model in which each core has a full copy of memory, its loads read from the local copy, and its writes update all copies according to some restrictions that define a model. Were TSO defined with this model, each store would write its own core's memory copy immediately and then possibly later update all other memories together.

Goodman [7] publicly discussed the idea of *processor consistency (PC)*, wherein a core's stores reach other cores in order but do not necessarily reach other cores at the same "time." Gharachorloo et al. [5] more precisely define PC. TSO and x86-TSO are special cases of PC in which each core sees its own store immediately, and when any other cores see a store, all other cores see it. This property is called *write atomicity* in the next chapter (Section 5.5).

To the best of our knowledge, TSO was first formally defined by Sindhu et al. [16]. As discussed, in Section 4.3, Sewell et al. [9, 14, 15] propose and formalize the x86-TSO model that appears consistent with AMD and Intel x86 documentation and current implementations.

4.6   COMPARING SC AND TSO

Now that we have seen two memory consistency models, we can compare them. How do SC, TSO, etc., relate?

- *Executions*: SC executions are a proper subset of TSO executions; all SC executions are TSO executions, while some TSO executions are SC executions and some are not. See the Venn diagram in Figure 4.5a.

- *Implementations*: Implementations follow the same rules: SC implementations are a proper subset of TSO implementations. See Figure 4.5b, which is the same as Figure 4.5a.

More generally, a memory consistency model Y is strictly more *relaxed* (*weaker*) than a memory consistency model X if all X executions are also Y executions, but not vice versa. If Y is more relaxed than X, then it follows that all X implementations are also Y implementations. It is also possible that two memory consistency models are incomparable because both allow executions precluded by the other.

As Figure 4.5 depicts, TSO is more relaxed than SC but less relaxed than incomparable models MC1 and MC2. In the next chapter, we will see candidates for MC1 and MC2, including a case study for the IBM Power memory consistency model.

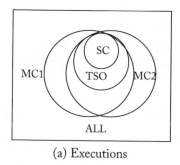
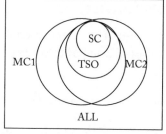

(a) Executions   (b) Implementations (same as (a))

Figure 4.5: Comparing memory consistency models.

What is a Good Memory Consistency Model?

A good memory consistency model should possess Sarita Adve's 3Ps [1] plus our fourth P:

- *Programmability*: A good model should make it (relatively) easy to write multithreaded programs. The model should be intuitive to most users, even those who have not read the details. It should be precise, so that experts can push the envelope of what is allowed.

- *Performance*: A good model should facilitate high-performance implementations at reasonable power, cost, etc. It should give implementors broad latitude in options.

- *Portability*: A good model would be adopted widely or at least provide backward compatibility or the ability to translate among models.

- *Precision*: A good model should be precisely defined, usually with mathematics. Natural languages are too ambiguous to enable experts to push the envelope of what is allowed.

How good are SC and TSO?

Using these 4Ps:

- *Programmability*: SC is the most intuitive. TSO is close because it acts like SC for common programming idioms. Nevertheless, subtle non-SC executions can bite programmers and tool authors.

- *Performance*: For simple cores, TSO can offer better performance than SC, but the difference can be made small with speculation.

- *Portability*: SC is widely understood, while TSO is widely adopted.

- *Precise*: SC and TSO are formally defined.

The bottom line is that SC and TSO are pretty close, especially compared with the more complex and more relaxed memory consistency models discussed in the next chapter.

4.7   REFERENCES

[1] S. V. Adve. Designing memory consistency models for shared-memory multiprocessors. Ph.D. thesis, Computer Sciences Department, University of Wisconsin–Madison, November 1993. 51

[2] W. W. Collier. *Reasoning About Parallel Architectures*. Prentice-Hall, Inc., 1990. 50

[3] Y. Duan, A. Muzahid, and J. Torrellas. WeeFence: Toward making fences free in TSO. In *The 40th Annual International Symposium on Computer Architecture*, 2013. DOI: 10.1145/2485922.2485941. 49

[4] Y. Duan, N. Honarmand, and J. Torrellas. Asymmetric memory fences: Optimizing both performance and implementability. In *Proc. of the 20th International Conference on Architectural Support for Programming Languages and Operating Systems*, 2015. DOI: 10.1145/2694344.2694388. 49

[5] K. Gharachorloo, D. Lenoski, J. Laudon, P. Gibbons, A. Gupta, and J. Hennessy. Memory consistency and event ordering in scalable shared-memory. In *Proc. of the 17th Annual International Symposium on Computer Architecture*, pp. 15–26, May 1990. DOI: 10.1109/isca.1990.134503. 50

[6] K. Gharachorloo, M. Sharma, S. Steely, and S. Van Doren. Architecture and design of AlphaServer GS320. In *Proc. of the 9th International Conference on Architectural Support for Programming Languages and Operating Systems*, 2000. DOI: 10.1145/378993.378997. 49

[7] J. R. Goodman. Cache consistency and sequential consistency. *Technical Report 1006*, Computer Sciences Department, University of Wisconsin–Madison, February 1991. 50

[8] C. Lin, V. Nagarajan, and R. Gupta. Efficient sequential consistency using conditional fences. In *19th International Conference on Parallel Architectures and Compilation Techniques*, 2010. DOI: 10.1145/1854273.1854312. 49

[9] S. Owens, S. Sarkar, and P. Sewell. A better x86 memory model: x86-TSO. In *Proc. of the Conference on Theorem Proving in Higher Order Logics*, 2009. DOI: 10.1007/978-3-642-03359-9_27. 46, 50

[10] B. Rajaram, V. Nagarajan, S. Sarkar, and M. Elver. Fast RMWs for TSO: Semantics and implementation. In *Proc. of ACM SIGPLAN Conference on Programming Language Design and Implementation*, 2013. DOI: 10.1145/2491956.2462196. 49

[11] A. Ros, T. E. Carlson, M. Alipour, and S. Kaxiras. Non-speculative load-load reordering in TSO. In *Proc. of the 44th Annual International Symposium on Computer Architecture*, 2017. DOI: 10.1145/3079856.3080220. 49

[12] A. Ros and S. Kaxiras. The superfluous load queue. In *51st Annual IEEE/ACM International Symposium on Microarchitecture*, 2018. DOI: 10.1109/micro.2018.00017. 49

[13] A. Ros and S. Kaxiras. Non-speculative store coalescing in total store order. In *45th ACM/IEEE Annual International Symposium on Computer Architecture*, 2018. DOI: 10.1109/isca.2018.00028. 49

[14] S. Sarkar, P. Sewell, F. Z. Nardelli, S. Owens, T. Ridge, T. Braibant, M. O. Myreen, and J. Alglave. The semantics of x86-CC multiprocessor machine code. In *Proc. of the 36th Annual ACM SIGPLAN-SIGACT Symposium on Principles of Programming Languages*, pp. 379–391, 2009. DOI: 10.1145/1480881.1480929. 46, 50

[15] P. Sewell, S. Sarkar, S. Owens, F. Z. Nardelli, and M. O. Myreen. x86-TSO: A rigorous and usable programmer's model for x86 multiprocessors. *Communications of the ACM*, July 2010. DOI: 10.1145/1785414.1785443. 45, 46, 50

[16] P. Sindhu, J.-M. Frailong, and M. Ceklov. Formal specification of memory models. *Technical Report CSL-91–11*, Xerox Palo Alto Research Center, December 1991. DOI: 10.1007/978-1-4615-3604-8_2. 50

CHAPTER 5

Relaxed Memory Consistency

The previous two chapters explored the memory consistency models *sequential consistency (SC)* and *total store order (TSO)*. These chapters presented SC as intuitive and TSO as widely implemented (e.g., in x86). Both models are sometimes called *strong* because the global memory order of each model usually respects (preserves) per-thread program order. Recall that SC preserves order for two memory operations from the same thread for all four combinations of loads and stores (Load → Load, Load → Store, Store → Store, and Store → Load), whereas TSO preserves the first three orders but not Store → Load order.

This chapter examines more *relaxed (weak) memory consistency models* that seek to preserve only the orders that programmers "require." The principal benefit of this approach is that mandating fewer ordering constraints can facilitate higher performance by allowing more hardware and software (compiler and runtime system) optimizations. The principal drawbacks are that relaxed models must formalize when ordering is "required" and provide mechanisms for programmers or low-level software to communicate such ordering to implementations, and vendors have failed to agree on a single relaxed model, compromising portability.

A full exploration of relaxed consistency models is beyond the scope of this chapter. This chapter is instead a primer that seeks to provide the basic intuition and help make the reader aware of the limits of a simple understanding of these models. In particular, we provide motivation for relaxed models (Section 5.1), present and formalize an example relaxed consistency model XC (Section 5.2), discuss the implementation of XC, including atomic instructions and instructions used to enforce ordering (Section 5.3), introduce sequential consistency for data-race-free programs (Section 5.4), present additional relaxed model concepts (Section 5.5), present the RISC-V and IBM Power memory model case studies (Section 5.6), point to further reading and other commercial models (Section 5.7), compare models (Section 5.8), and touch upon high-level-language memory models (Section 5.9).

5.1 MOTIVATION

As we will soon see, mastering relaxed consistency models can be more challenging than understanding SC and TSO. These drawbacks beg the question: why bother with relaxed models at all? In this section, we motivate relaxed models, first by showing some common situations in which programmers do not care about instruction ordering (Section 5.1.1) and then by discussing a few of the optimizations that can be exploited when unnecessary orderings are not enforced (Section 5.1.2).

5.1.1   OPPORTUNITIES TO REORDER MEMORY OPERATIONS

Consider the example depicted in Table 5.1. Most programmers would expect that r2 will always get the value NEW because S1 is before S3 and S3 is before the dynamic instance of L1 that loads the value SET, which is before L2. We can denote this:

- S1 → S3 → L1 loads SET → L2.

   Similarly, most programmers would expect that r3 will always get the value NEW because:

- S2 → S3 → L1 loads SET → L3.

   In addition to these two expected orders above, SC and TSO also require the orders S1 → S2 and L2 → L3. Preserving these additional orders may limit implementation optimizations to aid performance, yet these additional orders are not needed by the program for correct operation.

   Table 5.2 depicts a more general case of the handoff between two critical sections using the same lock. Assume that hardware supports lock acquire (e.g., with test-and-set doing a read-modify-write and looping until it succeeds) and lock release (e.g., store the value 0). Let core C1 acquire the lock, do critical section 1 with an arbitrary interleaving of loads (L1i) and stores (S1j), and then release the lock. Similarly, let core C2 do critical section 2, including an arbitrary interleaving of loads (L2i) and stores (S2j).

   Proper operation of the handoff from critical section 1 to critical section 2 depends on the order of these operations:

- All L1i, All S1j → R1 → A2 → All L2i, All S2j,

where commas (",") separate operations whose order is not specified.

   Proper operation does *not* depend on any ordering of the loads and stores within each critical section—unless the operations are to the same address (in which case ordering is required to maintain sequential processor ordering). That is:

- All L1i and S1j can be in any order with respect to each other, and

Table 5.1: What order ensures r2 and r3 always get NEW?

Core C1	Core C2	Comments
S1: data1 = NEW;		/* Initially, data1 and data2 = 0 & flag ≠ SET */
S2: data2 = NEW;		
S3: flag = SET;	L1: r1 = flag	/* spin loop: L1 & B1 may repeat many times */
	B1: if (r1 ≠ SET) goto L1;	
	L2: r2=data1;	
	L3: r3=data2;	

Table 5.2: What order ensures correct handoff from critical section 1 to 2?

Core C1	Core C2	Comments
A1: acquire(lock)		
/* Begin Critical Section 1 */		
Some loads L1i interleaved		/* Arbitrary interleaving of L1i's & S1j's */
with some stores S1j		
/* End Critical Section 1 */		/* Handoff from critical section 1 */
R1: release(lock)	A2: acquire(lock)	/* To critical section 2 */
	/* Begin Critical Section 2 */	
	Some loads L2i interleaved	/* Arbitrary interleaving of L2i's & S2j's */
	with some stores S2j	
	/* End Critical Section 2 */	
	R2: release(lock)	

- All L2i and S2j can be in any order with respect to each other.

If proper operation does *not* depend on ordering among many loads and stores, perhaps one could obtain higher performance by relaxing the order among them, since loads and stores are typically much more frequent than lock acquires and releases. This is what relaxed or weak models do.

5.1.2   OPPORTUNITIES TO EXPLOIT REORDERING

Assume for now a relaxed memory consistency model that allows us to reorder any memory operations unless there is a FENCE between them. This relaxed model forces the programmer to reason about which operations need to be ordered, which is a drawback, but it also enables many optimizations that can improve performance. We discuss a few common and important optimizations, but a deep treatment of this topic is beyond the scope of this primer.

Non-FIFO, Coalescing Write Buffer

Recall that TSO enables the use of a FIFO write buffer, which improves performance by hiding some or all of the latency of committed stores. Although a FIFO write buffer improves performance, an even more optimized design would use a non-FIFO write buffer that permits coalescing of writes (i.e., two stores that are not consecutive in program order can write to the same entry in the write buffer). However, a non-FIFO coalescing write buffer normally violates TSO because TSO requires stores to appear in program order. Our example relaxed model allows stores to coalesce in a non-FIFO write buffer, so long as the stores have not been separated by a FENCE.

Simpler Support for Core Speculation

In systems with strong consistency models, a core may speculatively execute loads out of program order before they are ready to be committed. Recall how the MIPS R10000 core, which supports SC, used such speculation to gain better performance than a naive implementation that did not speculate. The catch, however, is that speculative cores that support SC typically have to include mechanisms to check whether the speculation is correct, even if mis-speculations are rare [15, 21]. The R10000 checks speculation by comparing the addresses of evicted cache blocks against a list of the addresses that the core has speculatively loaded but not yet committed (i.e., the contents of the core's load queue). This mechanism adds to the cost and complexity of the hardware, it consumes additional power, and it represents another finite resource that may constrain instruction level parallelism. In a system with a relaxed memory consistency model, a core can execute loads out of program order without having to compare the addresses of these loads to the addresses of incoming coherence requests. These loads are not speculative with respect to the relaxed consistency model (although they may be speculative with respect to, say, a branch prediction or earlier stores by the same thread to the same address).

Coupling Consistency and Coherence

We previously advocated decoupling consistency and coherence to manage intellectual complexity. Alternatively, relaxed models can provide better performance than strong models by "opening the coherence box." For example, an implementation might allow a subset of cores to load the new value from a store even as the rest of the cores can still load the old value, temporarily breaking coherence's single-writer–multiple-reader invariant. This situation can occur, for example, when two thread contexts logically share a per-core write buffer or when two cores share an L1 data cache. However, "opening the coherence box" incurs considerable intellectual and verification complexity, bringing to mind the Greek myth about Pandora's box. As we will discuss in Section 5.6.2, IBM Power permits the above optimizations. We will also explore in Chapter 10 why and how GPUs and heterogeneous processors typically open the coherence box while enforcing consistency. But first we explore relaxed models with the coherence box sealed tightly.

5.2   AN EXAMPLE RELAXED CONSISTENCY MODEL (XC)

For teaching purposes, this section introduces an *eXample relaxed Consistency model (XC)* that captures the basic idea and some implementation potential of relaxed memory consistency models. XC assumes that a global memory order exists, as is true for the strong models of SC and TSO, as well as the largely defunct relaxed models for Alpha [33] and SPARC Relaxed Memory Order (RMO) [34].

5.2.1 THE BASIC IDEA OF THE XC MODEL

XC provides a FENCE instruction so that programmers can indicate when order is needed; otherwise, by default, loads and stores are unordered. Other relaxed consistency models call a FENCE a barrier, a memory barrier, a membar, or a sync. Let core Ci execute some loads and/or stores, Xi, then a FENCE instruction, and then some more loads and/or stores, Yi. The FENCE ensures that memory order will order all Xi operations before the FENCE, which in turn is before all Yi operations. A FENCE instruction does not specify an address. Two FENCEs by the same core also stay ordered. However, a FENCE does *not* affect the order of memory operations at *other* cores (which is why "fence" may be a better name than "barrier"). Some architectures include multiple FENCE instructions with different ordering properties; for example, an architecture could include a FENCE instruction that enforces all orders except from a store to a subsequent load. In this chapter, however, we consider only FENCEs that order all types of operations.

XC's memory order is guaranteed to respect (preserve) program order for:

- Load → FENCE

- Store → FENCE

- FENCE → FENCE

- FENCE → Load

- FENCE → Store

   XC maintains TSO rules for ordering two accesses to the *same* address only:

- Load → Load to same address

- Load → Store to the same address

- Store → Store to the same address

   These rules enforce the sequential processor model (i.e., sequential core semantics) and prohibit behaviors that might astonish programmers. For example, the Store → Store rule prevents a critical section that performs "A = 1" then "A = 2" from completing strangely with A set to 1. Likewise, the Load → Load rule ensures that if B was initially 0 and another thread performs "B = 1," then the present thread cannot perform "r1 = B" then "r2 = B" with r1 getting 1 and r2 getting 0, as if B's value went from new to old.

   XC ensures that loads immediately see updates due to their own stores (like TSO's write buffer bypassing). This rule preserves the sequentiality of single threads, also to avoid programmer astonishment.

5.2.2   EXAMPLES USING FENCES UNDER XC

Table 5.3 shows how programmers or low-level software should insert FENCEs in the program of Table 5.1 so that it operates correctly under XC. These FENCEs ensure:

- S1, S2 → F1 → S3 → L1 loads SET → F2 → L2, L3.

The F1 FENCE, which orders stores, makes sense to most readers, but some are surprised by the need for the F2 FENCE ordering loads. However, if one allows the loads to execute out of order, they can make it look like in-order stores executed out of order. For example, if execution can proceed as L2, S1, S2, S3, L1, and L3, then L2 can obtain the value 0. This result is especially possible for a program that does not include the B1 control dependence, so that L1 and L2 are consecutive loads to different addresses, wherein reordering seems reasonable, but is not.

Table 5.3: Adding FENCEs for XC to Table 5.1's program

Core C1	Core C2	Comments
S1: data1 = NEW;		/* Initially, data1 & data2 = 0 & flag ≠ SET */
S2: data2 = NEW;		
F1: FENCE		
S3: flag = SET;	L1: r1 = flag;	/* L1 & B1 may repeat many times */
	B1: if (r1≠ SET) goto L1;	
	F2: FENCE	
	L2: r2 = data1;	
	L3: r3 = data2;	

Table 5.4 shows how programmers or low-level software could insert FENCEs in the critical section program of Table 5.2 so that it operates correctly under XC. This FENCE insertion policy, in which FENCEs surround each lock acquire and lock release, is conservative for illustration purposes; we will later show that some of these FENCEs can be removed. In particular, FENCEs F13 and F22 ensure a correct handoff between critical sections because:

- All L1i, All S1j → **F13** → R11 → A21 → **F22** → All L2i, All S2j

Next, we formalize XC and then show why the above two examples work.

5.2.3   FORMALIZING XC

Here, we formalize XC in a manner consistent with the previous two chapters' notation and approach. Once again, let L(a) and S(a) represent a load and a store, respectively, to address a. Orders <p and <m define per-processor program order and global memory order, respectively.

Table 5.4: Adding FENCEs for XC to Table 5.2's critical section program. (Note that not all FENCES are strictly necessary for correctness.)

Core C1	Core C2	Comments
F11: FENCE		
A11: acquire(lock)		
F12: FENCE		
Some loads L1i interleaved		/* Arbitrary interleaving of L1i's & S1j's */
with some stores S1j		
F13: FENCE		
R11: release(lock)	**F21: FENCE**	/* Handoff from critical section 1 */
F14: FENCE	A21: acquire(lock)	/* To critical section 2 */
	F22: FENCE	
	Some loads L2i interleaved	/* Arbitrary interleaving of L2i's & S2j's */
	with some stores S2j	
	F23: FENCE	
	R22: release(lock)	
	F24: FENCE	

Program order <p is a per-processor total order that captures the order in which each core logically (sequentially) executes memory operations. Global memory order <m is a total order on the memory operations of all cores.

More formally, an **XC execution** requires the following.

1. All cores insert their loads, stores, and FENCEs into the order <m respecting:

    - If L(a) <p FENCE $\Rightarrow$ L(a) <m FENCE   /* Load $\rightarrow$ FENCE */
    - If S(a) <p FENCE $\Rightarrow$ S(a) <m FENCE   /* Store $\rightarrow$ FENCE */
    - If FENCE <p FENCE $\Rightarrow$ FENCE <m FENCE /* FENCE $\rightarrow$ FENCE */
    - If FENCE <p L(a) $\Rightarrow$ FENCE <m L(a)   /* FENCE $\rightarrow$ Load */
    - If FENCE <p S(a) $\Rightarrow$ FENCE <m S(a)   /* FENCE $\rightarrow$ Store */

2. All cores insert their loads and stores to the *same* address into the order <m respecting:

    - If L(a) <p L'(a) $\Rightarrow$ L(a) <m L' (a)   /* Load $\rightarrow$ Load to same address */
    - If L(a) <p S(a) $\Rightarrow$ L(a) <m S(a)   /* Load $\rightarrow$ Store to same address */
    - If S(a) <p S'(a) $\Rightarrow$ S(a) <m S' (a)   /* Store $\rightarrow$ Store to same address */

3. Every load gets its value from the last store before it to the same address:

Value of L(a) = Value of MAX $_{<m}$ {S(a) | S(a) <m L(a) or S(a) <p L(a)} /* Like TSO */

We summarize these ordering rules in Table 5.5. This table differs considerably from the analogous tables for SC and TSO. Visually, the table shows that ordering is enforced only for operations to the same address or if FENCEs are used. Like TSO, if operation 1 is "store C" and operation 2 is "load C," the store can enter the global order after the load, but the load must already see the newly stored value.

An implementation that allows only XC executions is an **XC implementation**.

Table 5.5: XC ordering rules. An "X" denotes an enforced ordering. A "B" denotes that bypassing is required if the operations are to the same address. An "A" denotes an ordering that is enforced only if the operations are to the same address. Entries different from TSO are shaded and shown in bold.

		Operation 2			
		Load	Store	RMW	FENCE
Operation 1	Load	**A**	**A**	**A**	X
	Store	B	**A**	**A**	X
	RMW	**A**	**A**	**A**	X
	FENCE	X	X	X	X

5.2.4   EXAMPLES SHOWING XC OPERATING CORRECTLY

With the formalisms of the last section, we can now reveal why Section 5.2.2's two examples work correctly. Figure 5.1a shows an XC execution of the example from Table 5.3 in which core C1's stores S1 and S2 are reordered, as are core C2's loads L2 and L3. Neither reordering, however, affects the results of the program. Thus, as far as the programmer can tell, this XC execution is equivalent to the SC execution depicted in Figure 5.1b, in which the two pairs of operations are not reordered.

Similarly, Figure 5.2a depicts an execution of the critical section example from Table 5.4 in which core C1's loads L1i and stores S1j are reordered with respect to each other, as are core C2's L2i and stores S2j. Once again, these reorderings do not affect the results of the program. Thus, as far as the programmer can tell, this XC execution is equivalent to the SC execution depicted in Figure 5.2b, in which no loads or stores are reordered.

These examples demonstrate that, with sufficient FENCEs, a relaxed model like XC can appear to programmers as SC. Section 5.4 discusses generalizing from these examples, but first let us implement XC.

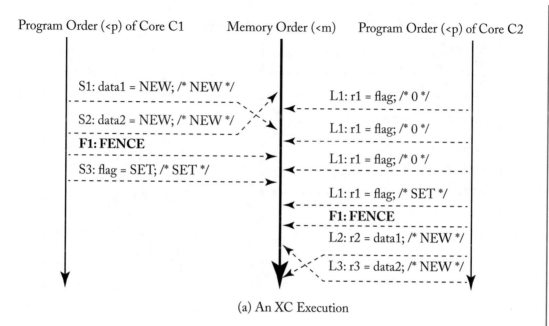

(a) An XC Execution

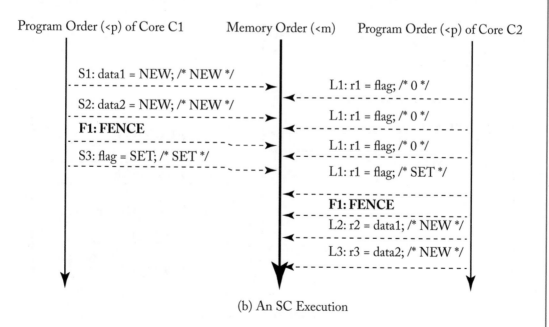

(b) An SC Execution

Figure 5.1: Two equivalent executions of Table 5.3's program.

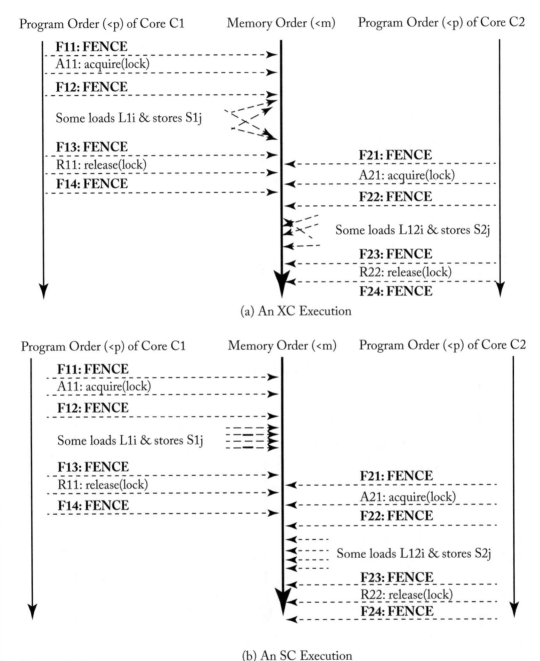

Figure 5.2: Two equivalent executions of Table 5.4's critical section program.

5.3   IMPLEMENTING XC

This section discusses implementing XC. We follow an approach similar to that used for implementing SC and TSO in the previous two chapters, in which we separate the reordering of core operations from cache coherence. Recall that each core in a TSO system was separated from shared memory by a FIFO write buffer. For XC, each core will be separated from memory by a more general *reorder unit* that can reorder both loads and stores.

As depicted by Figure 5.3a, XC operates as follows.

- Loads, stores, and FENCEs leave each core Ci in Ci's program order <p and enter the *tail* of Ci's reorder unit.

- Ci's reorder unit queues operations and passes them from its tail to its *head*, either in program order or reordered within by rules specified below. A FENCE gets discarded when it reaches the reorder unit's head.

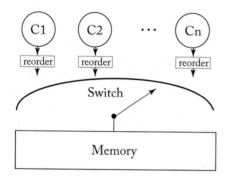

This XC implementation is modeled after the SC and TSO switch implementations of the previous chapter, except that a more-general reorder unit separates cores and the memory switch.

(a) An XC implementation using a switch

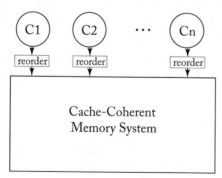

This XC implementation replaces the switch above with a cache-coherent memory system in a manner analogous to what was done for SC and TSO in the previous chapters.

(b) An XC implementation using cache coherence

Figure 5.3: Two XC implementations.

- When the switch selects core Ci, it performs the load or store at the head of Ci's reorder unit.

The reorder units obey rules for (1) FENCEs, (2) operations to the same address, and (3) bypassing.

1. FENCEs can be implemented in several different ways (see Section 5.3.2), but they must enforce order. Specifically, regardless of address, the reorder unit may *not* reorder:

   Load → FENCE, Store → FENCE, FENCE → FENCE, FENCE → Load, or FENCE → Store

2. For the *same* address, the reorder unit may *not* reorder:

   Load → Load, Load → Store, Store → Store (to the same address)

3. The reorder unit must ensure that loads immediately see updates due to their own stores.

Not surprisingly, all of these rules mirror those of Section 5.2.3.

In the previous two chapters, we argued that the switch and memory in SC and TSO implementations could be replaced by a cache-coherent memory system. The same argument holds for XC, as illustrated by Figure 5.3b. Thus, as for SC and TSO, an XC implementation can separate core (re)ordering rules from the implementation of cache coherence. As before, cache coherence implements the global memory order. What is new is that memory order can more often disrespect program order due to reorder unit reorderings.

So how much performance does moving from TSO to a relaxed model like XC afford? Unfortunately, the correct answer depends on the factors discussed in Section 5.1.2, such as FIFO vs. coalescing write buffers and speculation support.

In the late 1990s, one of us saw the trend toward speculative cores as diminishing the *raison d'être* for relaxed models (better performance) and argued for a return to the simpler interfaces of SC or TSO [22]. Although we still believe that simple interfaces are good, the move did not happen. One reason is corporate momentum. Another reason is that not all future cores will be highly speculative due to power-limited deployments in embedded chips and/or chips with many (asymmetric) cores.

5.3.1 ATOMIC INSTRUCTIONS WITH XC

There are several viable ways of implementing an atomic RMW instruction in a system that supports XC. The implementation of RMWs also depends on how the system implements XC; in this section, we assume that the XC system consists of dynamically scheduled cores, each of which is connected to the memory system by a non-FIFO coalescing write buffer.

In this XC system model, a simple and viable solution would be to borrow the implementation we used for TSO. Before executing an atomic instruction, the core drains the write buffer, obtains the block with read-write coherence permissions, and then performs the load part and the store part. Because the block is in a read-write state, the store part performs directly to the cache, bypassing the write buffer. Between when the load part performs and the store part performs, if such a window exists, the cache controller must not evict the block; if an incoming coherence request arrives, it must be deferred until the store part of the RMW performs.

Borrowing the TSO solution for implementing RMWs is simple, but it is overly conservative and sacrifices some performance. Notably, draining the write buffer is not required because XC allows both the load part and the store part of the RMW to pass earlier stores. Thus, it is sufficient to simply obtain read-write coherence permissions to the block and then perform the load part and the store part without relinquishing the block between those two operations.

Other implementations of atomic RMWs are possible, but they are beyond the scope of this primer. One important difference between XC and TSO is how atomic RMWs are used to achieve synchronization. In Table 5.6, we depict a typical critical section, including lock acquire and lock release, for both TSO and XC. With TSO, the atomic RMW is used to attempt to acquire the lock, and a store is used to release the lock. With XC, the situation is more complicated. For the acquire, XC does not, by default, constrain the RMW from being reordered with respect to the operations in the critical section. To avoid this situation, a lock acquire must be followed by a FENCE. Similarly, the lock release is not, by default, constrained from being reordered with respect to the operations before it in the critical section. To avoid this situation, a lock release must be preceded by a FENCE.

Table 5.6: Synchronization in TSO vs. synchronization in XC

Code	TSO	XC
Acquire lock	RMW: test-and-set L / * read L, write L=1 */ if L==1, goto RMW /* if lock held, try again */	RMW: test-and-set L / * read L, write L=1 */ if L==1, goto RMW /* if lock held, try again */ **FENCE**
Critical Section	Loads and stores	Loads and stores
Release lock	Store L=0	**FENCE** Store L=0

5.3.2   FENCES WITH XC

If a core C1 executes some memory operations Xi, then a FENCE, and then memory operations Yi, the XC implementation must preserve order. Specifically, the XC implementation must order Xi <m FENCE <m Yi. We see three basic approaches:

- An implementation can implement SC and treat all FENCEs as no-ops. This is not done (yet) in a commercial product, but there have been academic proposals, e.g., via implicit transactional memory [16].

- An implementation can wait for all memory operations Xi to perform, consider the FENCE done, and then begin memory operations Yi. This "FENCE as drain" method is common, but it makes FENCEs costly.

- An implementation can aggressively push toward what is necessary, enforcing Xi <m FENCE <m Yi, without draining. Exactly how one would do this is beyond the scope of this primer. While this approach may be more complex to design and verify, it can lead to better performance than draining.

In all cases, a FENCE implementation must know when each operation Xi is done (or at least ordered). Knowing when an operation is done can be especially tricky for a store that bypasses the usual cache coherence (e.g., a store to an I/O device or one using some fancy write update optimization).

5.3.3   A CAVEAT

Finally, an XC implementor might say, "I'm implementing a relaxed model, so anything goes." This is not true. One must obey the many XC rules, e.g., Load → Load order to the same address (this particular ordering is actually non-trivial to enforce in an out-of-order core). Moreover, all XC implementations must be strong enough to provide SC for programs that have FENCEs between each pair of instructions because these FENCEs require memory order to respect all of the program order.

5.4   SEQUENTIAL CONSISTENCY FOR DATA-RACE-FREE PROGRAMS

Children and computer architects would like "to have their cake and eat it too." For memory consistency models, this can mean enabling programmers to reason with the (relatively) intuitive model of SC while still achieving the performance benefits of executing on a relaxed model like XC.

Fortunately, simultaneously achieving both goals is possible for the important class of *data-race-free (DRF)* programs [3]. Informally, a data race occurs when two threads access the same memory location, at least one of the accesses is a write, and there are no intervening synchronization operations. Data races are often, but not always, the result of programming errors and many programmers seek to write DRF programs. *SC for DRF programming* asks programmers to ensure programs are DRF under SC by writing correctly synchronized programs and labeling the synchronization operations; it then asks implementors to ensure that all executions of DRF programs on the relaxed model are also SC executions by mapping the labeled synchronization operations to FENCEs and RMWs supported by the relaxed memory model. XC and most commercial relaxed memory models have the necessary FENCE instructions and RMWs to recover SC. Moreover, this approach also serves as a foundation for Java and C++ high-level language (HLL) memory models (Section 5.9).

Table 5.7: Example with four outcomes for XC with a data race

Core C1	Core C2	Comments
F11: FENCE		/* Initially, data1 & data2 = 0 */
A11: acquire(lock)		
F12: FENCE		
S1: data1 = NEW;	L1: r2 = data2;	
S2: data2 = NEW;	L2: r1 = data1;	/* Four Possible Outcomes under XC:
F13: FENCE		(r1, r2) =
R11: release(lock)		(0, 0), (0, NEW), (NEW, 0), or (NEW, NEW)
F14: FENCE		But has a Data Race */

Table 5.8: Example with two outcomes for XC without a data race, just like SC

Core C1	Core C2	Comments
F11: FENCE		/* Initially, data1 & data2 = 0 */
A11: acquire(lock)		
F12: FENCE		
S1: data1 = NEW;		
S2: data2 = NEW;		
F13: FENCE		
R11: release(lock)	**F21: FENCE**	
F14: FENCE	A21: acquire(lock)	
	F22: FENCE	
	L1: r2 = data2;	
	L2: r1 = data1;	/* Two Possible Outcomes under XC:
	F23: FENCE	(r1, r2) =
	R22: release(lock)	(0, 0) or (NEW, NEW)
	F24: FENCE	Same as with SC */

Let us motivate "SC for DRF" with two examples. Both Table 5.7 and Table 5.8 depict examples in which Core C1 stores two locations (S1 and S2) and Core C2 loads the two locations in the opposite order (L1 and L2). The examples differ because Core C2 does no synchronization in Table 5.7 but acquires the same lock as Core C1 in Table 5.8.

Since Core C2 does no synchronization in Table 5.7, its loads can execute concurrently with Core C1's stores. Since XC allows Core C1 to reorder stores S1 and S2 (or not) and Core C2 to reorder loads L1 and L2 (or not), four outcomes are possible wherein (r1, r2) = (0, 0), (0, NEW), (NEW, 0), or (NEW, NEW). Output (0, NEW) occurs, for example, if loads and stores execute in the sequence S2, L1, L2, and then S1 or the sequence L2, S1, S2, and then L1. However, this example includes two data races (S1 with L2 and S2 with L1) because Core C2 does not acquire the lock used by Core C1.

Table 5.8 depicts the case in which Core C2 acquires the same lock as acquired by Core C1. In this case, Core C1's critical section will execute completely before Core C2's or vice versa. This allows two outcomes: (r1, r2) = (0, 0) or (NEW, NEW). Importantly, these outcomes are not affected by whether, within their respective critical sections, Core C1 reorders stores S1 and S2 and/or Core C2 reorders loads L1 and L2. "A tree falls in the woods (reordered stores), but no one hears it (no concurrent loads)." Moreover, the XC outcomes are the same as would be allowed under SC. "SC for DRF" generalizes from these two examples to claim:

- either an execution has data races that expose XC's reordering of loads or stores, or

- the XC execution is data-race-free and indistinguishable from an SC execution.

A more concrete understanding of "SC for DRF" requires some definitions.

- Some memory operations are tagged as *synchronization* ("synchronization operations"), while the rest are tagged *data* by default ("data operations"). Synchronization operations include lock acquires and releases.

- Two data operations Di and Dj *conflict* if they are from different cores (threads) (i.e., not ordered by program order), access the same memory location, and at least one is a store.

- Two synchronization operations Si and Sj *conflict* if they are from different cores (threads), access the same memory location (e.g., the same lock), and at least one of the synchronization operations is a write (e.g., acquire and release of a spinlock conflict, whereas two read_locks on a reader-writer lock do not).

- Two synchronization operations Si and Sj *transitively conflict* if either Si and Sj conflict or if Si conflicts with some synchronization operation Sk, Sk <p Sk' (i.e., Sk is earlier than Sk' in a core K's program order), and Sk' transitively conflicts with Sj.

- Two data operations Di and Dj *race* if they conflict and they appear in the global memory order without an intervening pair of transitively conflicting synchronization operations by the same cores (threads) i and j. In other words, a pair of conflicting data operations Di <m Dj are *not* a data race if and only if there exists a pair of transitively conflicting synchronization operations Si and Sj such that Di <m Si <m Sj <m Dj.

- An SC execution is data-race-free (DRF) if no data operations race.

- A program is DRF if all its SC executions are DRF.

- A memory consistency model supports "SC for DRF programs" if all executions of all DRF programs are SC executions. This support usually requires some special actions for synchronization operations.

Consider the memory model XC. Require that the programmer or low-level software ensures that all synchronization operations are preceded and succeeded by FENCEs, as they are in Table 5.8.

With FENCEs around synchronization operations, XC supports SC for DRF programs. While a proof is beyond the scope of this work, the intuition behind this result follows from the examples in Table 5.7 and Table 5.8 discussed above.

Supporting SC for DRF programs allows many programmers to reason about their programs with SC and not the more complex rules of XC and, at the same time, benefit from any hardware performance improvements or simplifications XC enables over SC. The catch—and isn't there always a catch?—is that guaranteeing DRF at high performance (i.e., without labeling too many operations as synchronization) can be challenging.

- It is undecidable to determine *exactly* which memory operations can race and therefore must be tagged as synchronization. Figure 5.4 depicts an execution in which core C2's store should be tagged as synchronization—which determines whether FENCEs are actually necessary—only if one can determine whether C1's initial block of code does not halt, which is, of course, undecidable. Undecidability can be avoided by adding FENCEs whenever one is unsure whether a FENCE is needed. This is always correct, but may hurt performance. In the limit, one can surround all memory operations by FENCEs to ensure SC behavior for *any* program.

- Finally, programs may have data races that violate DRF due to bugs. The bad news is that, after data races, the execution may no longer obey SC, forcing programmers to reason about the underlying relaxed memory model (e.g., XC). The good news is that all executions will obey SC at least until the first data race, allowing some debugging with SC reasoning only [5].

In summary, a programmer using a relaxed memory system can reason about their program with two options:

- they can reason directly using the rules for what the model does and does not order (e.g., Table 5.5, etc.) or

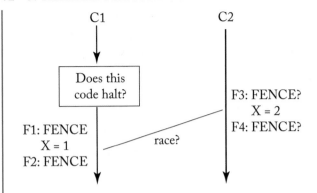

Core C2's FENCEs F3 and F4 are necessary only if core C1 executes "X = 1." Determining whether "X = 1" executes is undecidable, because it requires solving the undecidable halting problem.

Figure 5.4: Optimal placement of FENCEs is undecidable.

- they can insert sufficient synchronization to ensure no data races—synchronization races are still allowed—and reason about their program using the relatively simple model of sequential consistency that never appears to perform memory operations out of program order.

We recommend the latter "sequential consistency for data-race-free" approach almost always, leaving the former approach only to experts writing code like synchronization libraries or device drivers.

5.5   SOME RELAXED MODEL CONCEPTS

The academic literature offers many alternative relaxed memory models and concepts. Here we review some relaxed memory concepts from the vast literature to provide a basic understanding, but a full and formal exploration is beyond the scope of this primer. Fortunately, users of SC for DRF, which may be most programmers, do not have to master the concepts in this difficult section. On a first pass, readers may wish to skim or skip this section.

5.5.1   RELEASE CONSISTENCY

In the same ISCA 1990 session in which Adve and Hill proposed "SC for DRF," Gharachorloo et al. [20] proposed *release consistency (RC)*. Using the terminology of this chapter, the key observation of RC is that surrounding all synchronization operations with FENCEs is overkill. With a deeper understanding of synchronization, a synchronization *acquire* needs only a succeeding FENCE, while a synchronization *release* needs only a preceding FENCE.

For the critical section example of Table 5.4, FENCEs F11, F14, F21, and F24 may be omitted. Let us focus on "R11: release(lock)." FENCE F13 is important because it orders the critical section's loads (L1i) and stores (S1j) before the lock release. FENCE F14 may be omitted because there is no problem if core C1's subsequent memory operations (none shown in the table) were performed early before release R11.

RC actually allows these subsequent operations to be performed as early as the beginning of the critical section, in contrast to XC's FENCE, which disallows such an ordering. RC provides ACQUIRE and RELEASE operations that are similar to FENCEs, but order memory accesses in only one direction instead of in both directions like FENCEs. More generally, RC only requires:

ACQUIRE → Load, Store (but not Load, Store → ACQUIRE)
Load, Store → RELEASE (but not RELEASE → Load, Store) and
SC ordering of ACQUIREs and RELEASEs:
ACQUIRE → ACQUIRE
ACQUIRE → RELEASE
RELEASE → ACQUIRE, and
RELEASE → RELEASE

5.5.2 CAUSALITY AND WRITE ATOMICITY

Here we illustrate two subtle properties of relaxed models. The first property, *causality*, requires that, "If I see it and tell you about it, then you will see it too." For example, consider Table 5.9 where core C1 does a store S1 to update data1. Let core C2 spin until it sees the results of S1 (r1==NEW), perform FENCE F1, and then do S2 to update data2. Similarly, core C3 spins on load L2 until it sees the result of S2 (r2==NEW), performs FENCE F2, and then does L3 to observe store S1. If core C3 is guaranteed to observe S1 done (r3==NEW), then causality holds. On the other hand, if r3 is 0, causality is violated.

The second property, *write atomicity* (also called *store atomicity* or *multi-copy atomicity*), requires that a core's store is logically seen by all other cores at once. XC is write atomic by definition since its memory order (<m) specifies a logically atomic point at which a store takes effect at memory. Before this point, no *other* cores may see the newly stored value. After this point, all *other* cores must see the new value or the value from a later store, but not a value that was clobbered by the store. Write atomicity allows a core to see the value of its *own* store before it is seen by other cores, as required by XC, causing some to consider "write atomicity" to be a poor name.

A necessary, but not sufficient, condition for write atomicity is proper handling of the *Independent Read Independent Write (IRIW)* example. IRIW is depicted in Table 5.10 where cores C1 and C2 do stores S1 and S2, respectively. Assume that core C3's load L1 observes S1 (r1==NEW) and core C4's L3 observes S2 (r3==NEW). What if C3's L2 loads 0 (r2==0) and C4's L4 loads 0 (r4==0)? The former implies that core C3 sees store S1 before it sees S2, while the latter implies that C4 sees S2 before S1. In this case, stores S1 and S2 are not just "reordered," but *no order of stores even exists*, and write atomicity is violated. The converse is not necessarily true: proper handling of IRIW does not automatically imply store atomicity. Some more facts (that can make your head hurt and long for SC, TSO, or SC for DRF):

- Write atomicity implies causality. In Table 5.9, for example, core C2 observes store S1, performs a FENCE, and then does store S2. With write atomicity, this ensures C3 sees store S1 as done.

- Causality does *not* imply write atomicity. For Table 5.10, assume that cores C1 and C3 are two thread contexts of a multithreaded core that share a write buffer. Assume the same for cores C2 and C4. Let C1 put S1 in the C1–C3 write buffer, so it is observed by C3's L1 only. Similarly, C2 puts S2 into the C2–C4 write buffer, so S2 is observed by C4's L3 only. Let both C3 do L2 and C4 do L4 before either store leaves the write buffers. This execution violates write atomicity. Using the example in Table 5.9, however, one can see that this design provides causality.

Table 5.9: Causality: If I see a store and tell you about it, must you see it too?

Core C1	Core C2	Core C3
S1: data1 = NEW;		/* Initially, data1 & data2 = 0 */
	L1: r1 = data1;	
	B1: if (r ≠ NEW) goto L1;	
	F1: FENCE	
	S2: data2 = NEW;	
		L2: r2 = data2;
		B2: if (r2 ≠ NEW) goto L2;
		F2: FENCE
		L3: r3 = data1; /* r3==NEW? */

Table 5.10: **IRIW example: Must stores be in some order?**

Core C1	Core C2	Core C3	Core C4
S1: data1 = NEW;	S2: data2 = NEW;		/* Initially, data1=data2=0 */
		L1: r1 = data1; /* NEW */	
		F1: FENCE	L3: r3 = data2; /* NEW */
		L2: r2 = data2; /* NEW? */	**F2: FENCE**
			L4: r4 = data1; /* NEW? */

Finally, the XC memory model is both store atomic and maintains causality. We previously argued that XC was store atomic. XC maintains causality because store atomicity implies causality.

5.6   RELAXED MEMORY MODEL CASE STUDIES

In this section, we present two memory model case studies: RISC-V'S RVWMO (RISC-V Weak Memory Order) and IBM's Power Memory model.

5.6.1   RISC-V WEAK MEMORY ORDER (RVWMO)

RISC-V implements a memory model, RVWMO,[1] which can be understood as a mix of Release Consistency (RC) and XC. Similarly to XC, RVWMO is defined in terms of a global memory order (a total order of all memory operations) and has several variants of FENCE instructions. Similarly to RC, loads and stores can carry annotations: a load instruction can carry an ACQUIRE annotation, a store instruction can carry a RELEASE annotation, and an RMW instruction can be annotated with a RELEASE, ACQUIRE, or both RELEASE and ACQUIRE. In the following, we will provide a summary of the RVWMO memory model, explaining how RVWMO combines aspects of XC and RC. We will also discuss how RVWMO is subtly stronger than XC in some aspects while being weaker in other aspects.

RELEASE/ACQUIRE orderings. There are two types of ACQUIRE annotations: ACQUIRE-RC_{PC} and ACQUIRE-RC_{SC}.[2] Likewise, there are two types of RELEASE annotations: RELEASE-RC_{PC} and RELEASE-RC_{SC}. Whereas a load (store) can carry either of the ACQUIRE (RELEASE) annotations, an RMW can carry only RC_{SC} annotations. The annotations preserve the following orderings:

- ACQUIRE → Load,Store (ACQUIRE refers to both ACQUIRE-RC_{SC} and ACQUIRE-RC_{PC})

- Load,Store → RELEASE (RELEASE refers to both RELEASE-RC_{SC} and RELEASE-RC_{PC})

- RELEASE-RC_{SC} → ACQUIRE-RC_{SC}

FENCE orderings. RVWMO has several variants of FENCE instructions. There is a strong FENCE instruction, FENCE RW,RW,[3] which, like the one in XC, enforces the Load,Store → Load,Store orderings. In addition, there are five other non-trivial combinations: FENCE RW, W; FENCE R,RW; FENCE R,R; FENCE W,W; and FENCE.TSO. FENCE R,R, for

---

[1]RISC-V also specifies a TSO variant called zTSO which is not discussed in this section.
[2]PC for Processor Consistency and SC for Sequential Consistency. PC (Section 4.5) is a less-formal precursor to TSO, without TSO's write atomicity property.
[3]R denotes a read (load) and W denotes a write (store).

instance, enforces the Load → Load orderings only. FENCE.TSO enforces the Load → Load, Store → Store and the Load → Store orderings but not Store → Load.

Table 5.11: RVWMO: FENCE and RELEASE-ACQUIRE orderings ensure that both r1 and r2 cannot read 0

Core C1	Core C2	Comments
S1: RELEASE-RC$_{SC}$ x = NEW; L1: ACQUIRE-RC$_{SC}$ r1 = y;	S2: y = NEW; FENCE RW, RW; L2: r2=x;	Initially x=y=0

An Example. Table 5.11 shows an example with both RELEASE-ACQUIRE and FENCE orderings. S1 → L1 is enforced in core C1 due to the former and S2 → L2 is enforced in core C2 due to the latter. The combination, thus, ensures that both r1 and r2 cannot read 0.

Table 5.12: Address dependency induced ordering. Can r1=&data2 and r2=0?

Core C1	Core C2	Comments
S1: data2 = NEW; FENCE W,W S2: pointer = &data2;	L1: r1 = pointer; L2: r2=*r1;	Initially pointer = & data1, data1 = 0

Table 5.13: Data dependency induced ordering. Can r1 and r2 read 42?

Core C1	Core C2	Comments
L1: r1 = x; S1: y = r1;	L2: r2 = y; S2: x = r2;	Initially x=y=0

Dependency-induced orderings. RVWMO is subtly stronger than XC in some respects. Address, data, and control dependencies can induce memory orderings in RVWMO but not in XC. Consider the example shown in Table 5.12. Here, core C1 writes NEW to data2 and then sets pointer to point to the location of data2. (The two stores S1 and S2 are ordered via a FENCE W,W instruction). In core C2, L1 loads the value of pointer into r1 and then load L2 dereferences r1. Despite the two loads L1 and L2 not being explicitly ordered, RVWMO implicitly enforces L1 → L2 because there is an address dependency between L1 and L2: the value produced by L1 is dereferenced by L2.

Consider the example shown in Table 5.13, also referred to as *load buffering*. Let us assume that both x and y are initially 0. Can both r1 and r2 be allowed to read an arbitrary value (say

42), *out of thin air*? Somewhat surprisingly, XC does not prohibit this behavior. Because there is no FENCE between either L1 and S1 nor L2 and S2, XC does not enforce either of the Load → Store orderings. This can cause an execution in which:

- S1 predicts that L1 will read 42 and then speculatively writes 42 to y,

- L2 reads 42 from y into r2,

- S2 writes 42 to x, and

- L1 reads 42 from x into r1, thereby making the initial prediction "correct."

RVWMO, however, disallows this behavior by implicitly mandating both Load→Store orderings (L1 → S1 and L2 → S2) since there is a data dependency between each of the loads and the stores: the value read by each of the loads is written by the following store.

   In a similar vein, RVWMO also implicitly enforces an ordering between a load and a subsequent store that is control dependent on the load. This is to prevent a causality cycle, for example shown in Table 5.14, wherein a store is able to affect the value read by the preceding load which decides whether or not the store must execute.

Table 5.14: Control dependency induced ordering. Can r1 read 42, causing S1 (indirectly) to affect its own execution?

Core C1	Core C2	Comments
L1: r1=x; B1: if r1 ≠ 42 return; S1: y=42;	L2: r2 = y; S2: x= r2;	Initially x = y = 0

   It is worth noting that all of the above dependencies refer to syntactic dependencies rather than semantic dependencies, i.e, whether or not there is any dependency is a function of the identities of the registers, rather than the actual values. In addition to address, data and control, RVWMO also enforces "pipeline dependencies" to reflect behaviors of most processor pipeline implementations; a discussion of this is beyond the scope of this primer and readers are referred to the RISC-V specification [31].

Same address orderings. Recall that XC maintains TSO ordering rules for accesses to the same address. Like XC, RVWMO also enforces Load → Store, Store → Store orderings and does not enforce Store → Load orderings to the same address. Differently from XC, RVWMO does not enforce Load → Load ordering to the same address *in all situations*. It is only enforced when: (a) there is no store to the same address between the two loads and (b) the two loads return values written by different stores. For a detailed discussion on the rationale for this subtlety, readers are referred to the RISC-V specification [31].

RMWs. RISC-V supports two types of RMWs: atomic memory operations (AMO) and load reserved/store conditionals (LdR/StC). Whereas AMOs come from a single instruction (e.g., fetch-and-increment), LdR/StCs are actually composed of two separate instructions: the LdR which brings in a value and a reservation to the core and StC which succeeds only if the reservation is still held. The atomicity semantics of the two types of RMWs are subtly different. Similarly to an XC RMW, an AMO is said to be atomic if the load and store appear consecutively in the global memory order. LdR/StC is weaker. Suppose the LdR reads a value produced by store s; an LdR/StC is said to be atomic as long as there are no stores to the same address between s and the StC in the global memory order.

Summary. In summary, RVWMO is a recent relaxed memory model that combines aspects of XC and RC. For a detailed prose and formal specification, readers are referred to the RISC-V specification [31].

5.6.2   IBM POWER

IBM Power implements the Power memory model [23] (see especially Book II's Chapter 1, Section 4.4, and Appendix B). We attempt to give the gist of the Power memory model here, but we refer the reader to the Power manual for the definitive presentation, especially for programming Power. We do not provide an ordering table like Table 5.5 for SC because we are not confident we could specify all entries correctly. We discuss normal cacheable memory only ("Memory Coherence" enabled, "Write Through Required" disabled, and "Caching Inhibited" disabled) and not I/O space, etc. PowerPC [24] represents earlier versions of the current Power model. On a first pass of this primer, readers may wish to skim or skip this section; this memory model is significantly more complicated than the models presented thus far in this primer.

Power provides a relaxed model that is superficially similar to XC but with important differences that include the following.

First, stores in Power are performed *with respect to (w.r.t.) other cores*, not w.r.t. memory. A store by core C1 is "performed w.r.t." core C2 when any loads by core C2 to the same address will see the newly stored value or a value from a later store, but not the previous value that was clobbered by the store. Power ensures that if core C1 uses FENCEs to order store S1 before S2 and before S3, then the three stores will be performed w.r.t. every other core Ci in the same order. In the absence of FENCEs, however, core C1's store S1 may be performed w.r.t. core C2 but not yet performed w.r.t. to C3. Thus, Power is not guaranteed to create a total memory order (<m) as did XC.

Second, some FENCEs in Power are defined to be *cumulative*. Let a core C2 execute some memory accesses X1, X2, . . . , a FENCE, and then some memory accesses Y1, Y2,.... Let set X = {Xi} and set Y = {Yi}. (The Power manual calls these sets A and B, respectively.) Define *cumulative* to mean three things: (a) add to set X the memory accesses by *other* cores that are ordered before the FENCE (e.g., add core C1's store S1 to X if S1 is performed w.r.t. core C2 before C2's FENCE); (b) add to set Y the memory accesses by *other* cores that are ordered after the FENCE by data dependence, control dependence, or another FENCE; and (c) apply (a) recursively backward (e.g., for cores that have accesses previously ordered with core C1) and apply (b) recursively forward. (FENCEs in XC are also cumulative, but their cumulative behavior is automatically provided by XC's total memory order, not by the FENCEs specifically.)

Third, Power has three kinds of FENCEs (and more for I/O memory), whereas XC has only one FENCE.

- SYNC or HWSYNC ("HW" means "heavy weight" and "SYNC" stands for "synchronization") orders all accesses X before all accesses Y and is cumulative.

- LWSYNC ("LW" means "light weight") orders loads in X before loads in Y, orders loads in X before stores in Y, and orders stores in X before stores in Y. LWSYNC is cumulative. Note that LWSYNC does *not* order stores in X before loads in Y.

- ISYNC ("I" means "instruction") is sometimes used to order two loads from the same core, but it is not cumulative and, despite its name, it is not a FENCE like HWSYNC and LWSYNC, because it orders instructions and not memory accesses. For these reasons, we do not use ISYNC in our examples.

Fourth, Power orders accesses in some cases even without FENCEs. For example, if load L1 obtains a value used to calculate an effective address of a subsequent load L2, then Power orders load L1 before load L2. Also, if load L1 obtains a value used to calculate an effective address or data value of a subsequent store S2, then Power orders load L1 before store S2.

Table 5.15 illustrates Power's LWSYNC in action. Core C1 executes a LWSYNC to order data stores S1 and S2 before S3. Note that the LWSYNC does not order stores S1 and S2 with respect to each other, but this is not needed. A LWSYNC provides sufficient order here because it orders stores in X (S1 and S2) before stores in Y (S3). Similarly, core C2 executes a LWSYNC after its conditional branch B1 to ensure that load L1 completes with r1 assigned to SET before loads L2 and L3 execute. HWSYNC is not required because neither core needs to order stores before loads.

Table 5.16 illustrates Power's HWSYNC in action on a key part of Dekker's algorithm. The HWSYNCs ensure core C1's store S1 is before load L1 and core C2's store S2 is before load L2. This prevents the execution from terminating with r1=0 and r2=0. Using LWSYNC would not prevent this execution because an LWSYNC does not order earlier stores before later loads.

As depicted in Table 5.17, Power's LWSYNCs can be used to make the causality example from Table 5.9 behave sensibly (i.e., r3 is always set to NEW). LWSYNC F1 is executed only after load L1 sees the new value of data1, which means that store S1 has been performed w.r.t. core C2. LWSYNC F1 orders S1 before S2 with respect to core C2 by the cumulative property. LWSYNC F2 is executed only after load L2 sees the new value of data2, which means that store S2 has been performed w.r.t. core C3. The cumulative property also ensures that store S1 has been performed w.r.t core C3 before S2 (because of LWSYNC F1). Finally, LWSYNC F2 orders load L2 before L3, ensuring r3 obtains the value NEW.

Table 5.15: Power LWSYNCs to ensure r2 and r3 always get NEW

Core C1	Core C2	Comments
S1: data1 = NEW;		/* Initially, data1 & data2 = 0 & flag ≠ SET */
S2: data2 = NEW;		
F1: LWSYNC		/* Ensures S1 and S2 befor S3 */
S3: flag = SET;	L1: r1 = flag;	/* spin loop: L1 & B1 may repeat many times */
	B1: if (r1 ≠ SET) goto L1;	
	F2: LWSYNC	/* Ensure B1 before L2 and L3 */
	L2: r2 = data1;	
	L3: r3 = data2;	

Table 5.16: Power HWSYNCs to ensure both r1 and r2 are not set to 0

Core C1	Core C2	Comments
S1: x = NEW;	S2: y = NEW;	/* Initially, x = 0 & y = 0 */
F1: HWSYNC	**F2: HWSYNC**	/* Ensure S1 before Li for i = 1,2 */
L1: r1 = y;	L2: r2 = x;	

Table 5.17: Power LWSYNCs to ensure causality (r3 == NEW)

Core C1	Core C2	Core C3
S1: data1 = NEW;		/* Initially, data1 & data2 = 0 */
	L1: r1 = data1;	
	B1: if (r1 ≠ NEW) goto L1;	
	F1: LWSYNC	
	S2: data2 = NEW;	
		L2: r2 = data2;
		B2: if (r2 ≠ NEW) goto L2;
		F2: LWSYNC
		L3: r3 = data1; /* r3==NEW? */

As depicted in Table 5.18, Power's HWSYNCs can be used to make the Independent Read Independent Write Example (IRIW) of Table 5.10 behave sensibly (i.e., disallowing the result r1==NEW, r2==0, r3==NEW and r4==0). Using LWSYNCs is not sufficient. For example, core C3's F1 HWSYNC must cumulatively order core C1's store S1 before core C3's load L2.

An alternative way to look at the Power memory model is to specify the FENCEs needed to make Power behave like SC. Power can be restricted to SC executions by inserting an HWSYNC between each pair of memory-accessing instructions.[4] The above is a thought experiment, and it is definitely not a recommended way to achieve good performance on Power.

Table 5.18: Power HWSYNCs with the IRIW example

Core C1	Core C2	Core C3	Core C4
S1: data1 = NEW;	S2: data2 = NEW;		/* Initially, data1 = data2 = 0 */
		L1: r1 = data1; /* NEW */	L3: r3 = data2; /* NEW */
		F1: HWSYNC	**F2: HWSYNC**
		L2: r2 = data2; /* NEW? */	L4: r4 = data1; /* NEW? */

5.7   FURTHER READING AND COMMERCIAL RELAXED MEMORY MODELS

5.7.1   ACADEMIC LITERATURE

Below are a few highlights from the vast relaxed memory consistency literature. Among the first developed relaxed models was that of Dubois et al. [18] with *weak ordering*. Adve and Hill generalized weak ordering to the order strictly necessary for programmers with "SC for DRF" [3, 4]. Gharachorloo et al. [20] developed release consistency, as well as "proper labeling" (that can be viewed as a generalization of "SC for DRF") and a model that allows synchronization operations to follow TSO ("RC$_{PC}$"). Adve and Gharachorloo [2] wrote a seminal memory model tutorial summarizing the state of the art in the mid-1990s.

As far as we know, Meixner and Sorin [29] were the first to prove correctness of a relaxed memory model realized by separating the cores and the cache-coherent memory system with a reorder unit governed by certain rules.

5.7.2   COMMERCIAL MODELS

In addition to Power [23], commercial relaxed memory models include Alpha [33], SPARC RMO [34], and ARM [9, 10].

Alpha [33] is largely defunct, but, like XC, assumes a total memory order. Alpha retains some importance because it had a large influence on Linux synchronization since Linux ran on Alpha [28] (see especially Chapter 12 and Appendix C). McKenney [28] points out that Alpha

---

[4]FENCEs cannot restore SC on IBM Power when multiple access sizes are used [19].

did not order two loads even if the first provided the effective address for the second. More generally, McKenney's online book is a good source of information on Linux synchronization and its interaction with memory consistency models. Alglave et al. [8] provide a formal and an executable model for the same.

SPARC Relaxed Memory Order (RMO) [34] also provides a total memory order like XC. Although SPARC allows the operating system to select a memory model among TSO, PSO, and RMO, all current SPARC implementations operate with TSO in all cases. TSO is a valid implementation of PSO and RMO since it is strictly stronger.

ARMv7 [9, 10] provides a memory model similar in spirit to IBM Power. Like Power, it appears to not guarantee a total memory order. Like Power, ARM has multiple flavors of FENCEs, including a data memory barrier that can order all memory accesses or just stores and an instruction synchronization barrier like Power's ISYNC, as well as other FENCEs for I/O operations. For ARMv8 [11], ARM switched to a multi-copy atomic memory model that provides a total memory order. The model is similar in spirit to RISC-V's RVWMO,[5] allowing for loads and stores to be annotated with ACQUIRE and RELEASE annotations.

5.8 COMPARING MEMORY MODELS

5.8.1 HOW DO RELAXED MEMORY MODELS RELATE TO EACH OTHER AND TSO AND SC?

Recall that a memory consistency model Y is strictly more *relaxed* (*weaker*) than a memory consistency model X if all X executions (implementations) are also Y executions (implementations), but not vice versa. It is also possible that two memory consistency models are incomparable because both allow executions (implementations) precluded by the other.

Figure 5.5 repeats a figure from the previous chapter where Power replaces the previously unspecified MC1, while MC2 could be Alpha, ARM, RMO, or XC. How do they compare?

• Power is more relaxed than TSO which is more relaxed than SC.

• Alpha, ARM, RMO, and XC are more relaxed than TSO which is more relaxed than SC.

• Power is assumed to be incomparable with respect to Alpha, ARM, RMO, and XC until someone proves that one is more relaxed than the other or that the two are equivalent.

Mador-Haim et al. [26] developed an automated technique for comparing memory consistency models—including SC, TSO, and RMO—but they did not consider ARM or Power. ARM and Power may be equivalent, but we await a proof.

5.8.2 HOW GOOD ARE RELAXED MODELS?

As discussed in the previous chapter, a good memory consistency model should possess Sarita Adve's 3Ps plus our fourth P.

---

[5]but was defined before RVWMO was specified

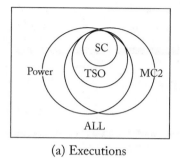

(a) Executions

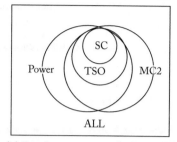

(b) Implementations (same as (a))

Figure 5.5: Comparing memory consistency models.

- *Programmability*: Relaxed model programmability is acceptable for those who use "SC for DRF." Deeply understanding a relaxed model (e.g., to write compilers and runtime systems) is difficult.

- *Performance*: Relaxed memory models can offer better performance than TSO, but the difference is smaller for many core micro-architectures.

- *Portability*: Porting while staying within "SC for DRF" is manageable. Pushing the limits of relaxed models, especially those that are incomparable, is hard.

- *Precise*: Many relaxed models are only informally defined. Moreover, formal definitions of relaxed models tend to be obtuse.

The bottom line is that … there is no simple bottom line.

5.9 HIGH-LEVEL LANGUAGE MODELS

The previous two chapters and this chapter so far address memory consistency models at the interface between hardware and low-level software, discussing (a) what software authors should expect and (b) what hardware implementors may do.

It is also important to define memory models for high-level languages (HLLs), specifying (a) what HLL software authors should expect and (b) what implementors of compilers, runtime systems, and hardware may do. Figure 5.6 illustrates the difference between (a) high-level and (b) low-level memory models.

Because many HLLs emerged in a largely single-threaded world, their specifications omitted a memory model. (Do you remember one in Kernighan and Ritchie [25]?) Java was perhaps the first mainstream language to have a memory model, but the first version had some issues [30].

Recently, however, memory models have been re-specified for Java [27] and specified for C++ [14]. Fortunately, the cornerstone of both models is a familiar one—SC for DRF [3]—provided in part by Sarita Adve's coauthorship of all three papers. To allow synchronization races

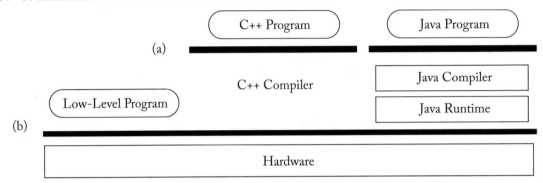

Figure 5.6: (a) High-level language (HLL) and (b) hardware memory models.

but not data races, programmers must tag variables as synchronization when it is possible they can race, using keywords such as **atomic**, or create synchronization locks implicitly with Java's monitor-like **synchronized** methods. In all cases, implementations are free to reorder references as long as data-race-free programs obey SC.

In particular, implementations can reorder or eliminate memory accesses between synchronization accesses. Figure 5.7 illustrates an example. Figure 5.7(a) presents the HLL code, and Figure 5.7(b) and (c) show executions on core C1 both without register allocation and with variable A allocated in register r1 so that load L3 is reordered and merged with load L1. This reordering is correct because, with "SC for DRF," no other thread "can be looking." In addition to register allocation, many, if not most, compiler and runtime optimizations—such as constant propagation, common subexpression elimination, loop fission/fusion, loop invariant code motion, software pipelining, and instruction scheduling—can reorder memory accesses. Because "SC for DRF" allows these optimizations, compilers and runtime systems can produce code with performance that is comparable to single-threaded code.

What if an HLL program has data races by accident or on purpose? Or what if expert programmers want to impose minimal ordering (or even no ordering) on synchronization operations, exploiting application knowledge? In this case, another thread could observe a non-SC execution. For Figure 5.7, core C2 (not shown) could update A and C and have core C1 observe the update to C but not A, which has been register allocated. This execution violates SC at the HLL level.

More generally, what are the limits on what threads can observe with data races or with non-SC synchronization operations? Specifically, Java requires security guarantees for all programs; C++, for performance considerations, supports a variety of synchronization operations (dubbed low-level atomics) that are weaker than SC. For these reasons, Java and C++ must specify behavior in all cases with the following goals:

1. allow all optimizations for high-performance DRF programs,

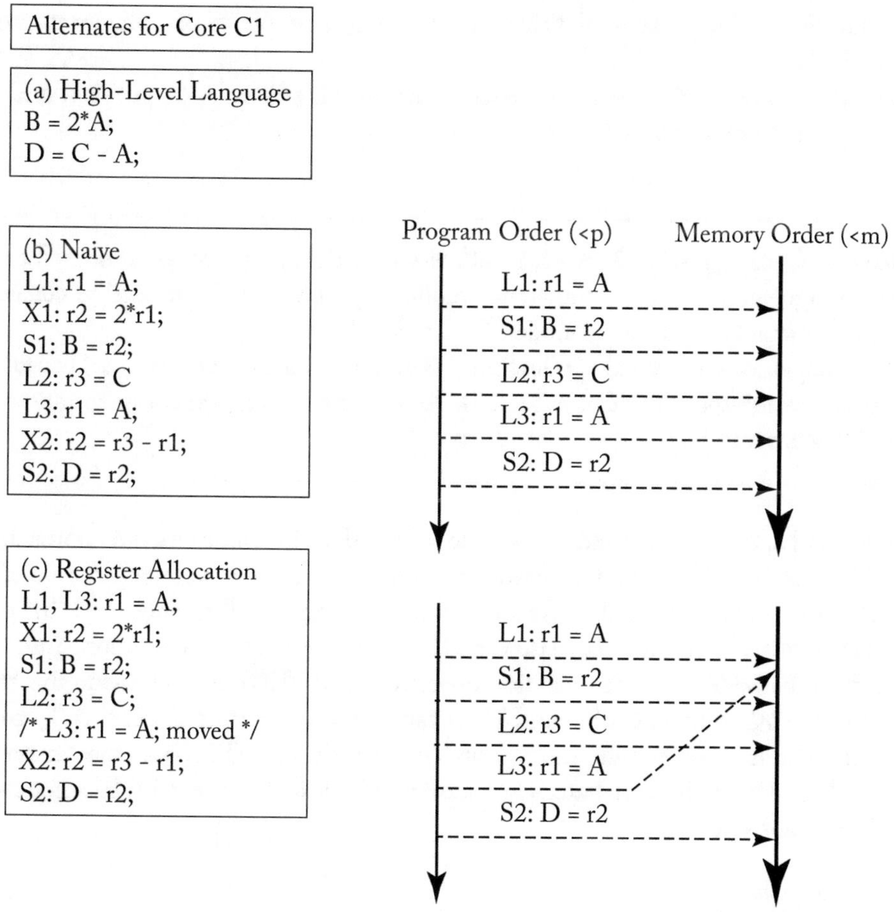

```
┌─────────────────────────────┐
│  Alternates for Core C1     │
└─────────────────────────────┘

┌─────────────────────────────┐
│  (a) High-Level Language    │
│  B = 2*A;                   │
│  D = C - A;                 │
└─────────────────────────────┘
```

Program Order (<p)   Memory Order (<m)

```
┌─────────────────────────────┐
│  (b) Naive                  │       L1: r1 = A
│  L1: r1 = A;                │       S1: B = r2
│  X1: r2 = 2*r1;             │       L2: r3 = C
│  S1: B = r2;                │       L3: r1 = A
│  L2: r3 = C                 │       S2: D = r2
│  L3: r1 = A;                │
│  X2: r2 = r3 - r1;          │
│  S2: D = r2;                │
└─────────────────────────────┘
```

```
┌─────────────────────────────┐
│  (c) Register Allocation    │       L1: r1 = A
│  L1, L3: r1 = A;            │       S1: B = r2
│  X1: r2 = 2*r1;             │       L2: r3 = C
│  S1: B = r2;                │       L3: r1 = A
│  L2: r3 = C;                │       S2: D = r2
│  /* L3: r1 = A; moved */    │
│  X2: r2 = r3 - r1;          │
│  S2: D = r2;                │
└─────────────────────────────┘
```

Figure 5.7: Register allocation affects memory order.

2. unambiguously, specify the allowed behavior of all programs, and

3. make this unambiguous specification simple.

In our judgment, the 2005 Java memory model made substantial progress, arguably succeeding on (2), mostly succeeding on (1)—it disallowed some compiler optimizations [32]—but not on (3). Fortunately, most programmers can use "SC for DRF" and not suffer the consequences of the "dark corners" of the Java memory model. Authors of compilers and runtime software, however, must confront them at some level. Consider this "dark corner" that must be understood: regardless of optimizations, a load to address A must always return a value stored to address A at some time (perhaps at initialization) and not a value "out of thin air" as illustrated in Table 5.13. Unfortunately, this is not the only example of complexity that must be confronted [12].

Satisfying all of the three requirements from above is still an open problem and actively researched in the context of C++ [13, 17].

Flashback to Quiz Question 5: A programmer who writes properly synchronized code relative to the high-level language consistency model (e.g,. Java) does not need to consider the architecture's memory consistency model. *True or false?*
Answer: It depends. For typical application programmers, the answer is *True*, because their programs behave as expected (SC is expected). For the programmers of compilers and runtime systems, the answer is *False*.

Thus, HLLs, such as Java and C++, adopt the relaxed memory model approach of "SC for DRF." When these HLLs run on hardware, should the hardware's memory model also be relaxed? On one hand, (a) relaxed hardware can give the most performance, and (b) compilers and runtime software need only translate the HLL's synchronizations operations into the particular hardware's low-level synchronization operations and FENCEs to provide the necessary ordering. On the other hand, (a) SC and TSO can give good performance, and (b) compilers and runtime software can generate more portable code without FENCEs from incomparable memory models. Although the debate is not settled, it is clear that relaxed HLL models do not *require* relaxed hardware.

5.10 REFERENCES

[1] (Don't) Do It Yourself: Weak Memory Models. http://diy.inria.fr/

[2] S. V. Adve and K. Gharachorloo. Shared memory consistency models: A tutorial. *IEEE Computer*, 29(12):66–76, December 1996. DOI: 10.1109/2.546611. 81

[3] S. V. Adve and M. D. Hill. Weak ordering—a new definition. In *Proc. of the 17th Annual International Symposium on Computer Architecture*, pp. 2–14, May 1990. DOI: 10.1109/isca.1990.134502. 68, 81, 83

[4] S. V. Adve and M. D. Hill. A unified formalization of four shared-memory models. *IEEE Transactions on Parallel and Distributed Systems*, June 1993. DOI: 10.1109/71.242161. 81

[5] S. V. Adve, M. D. Hill, B. P. Miller, and R. H. B. Netzer. Detecting data races on weak memory systems. In *Proc. of the 18th Annual International Symposium on Computer Architecture*, pp. 234–43, May 1991. DOI: 10.1145/115952.115976. 71

[6] J. Alglave, L. Maranget, S. Sarkar, and P. Sewell. Fences in weak memory models. In *Proc. of the International Conference on Computer Aided Verification*, July 2010. DOI: 10.1007/978-3-642-14295-6_25.

[7] J. Alglave, L. Maranget, S. Sarkar, and P. Sewell. Litmus: Running tests against hardware. In *Proc. of the International Conference on Tools and Algorithms for the Construction and Analysis of Systems*, March 2011. DOI: 10.1007/978-3-642-19835-9_5.

[8] J. Alglave, L. Maranget, P.E. McKenney, A. Parri, and A. Stern. Frightening small children and disconcerting grown-ups: Concurrency in the Linux kernel. In *ASPLOS*, 2018. DOI: 10.1145/3296957.3177156. 82

[9] ARM. *ARM Architecture Reference Manual, ARMv7-A and ARMv7-R Edition Errata Markup*. Downloaded January 13, 2011. 81, 82

[10] ARM. *ARM v7A+R Architectural Reference Manual*. Available from ARM Ltd. 81, 82

[11] ARM. *ARM v8 Architectural Reference Manual*. Available from ARM Ltd. 82

[12] M. Batty, K. Memarian, K. Nienhuis, J. Pichon-Pharabod, and P. Sewell. The problem of programming language concurrency semantics. In *ESOP*, 2015. DOI: 10.1007/978-3-662-46669-8_12. 86

[13] M. Batty, S. Owens, S. Sarkar, P. Sewell, and T. Weber. Mathematizing C++ concurrency. In *POPL*, 2011. DOI: 10.1145/1926385.1926394. 86

[14] H.-J. Boehm and S. V. Adve. Foundations of the C++ concurrency memory model. In *Proc. of the Conference on Programming Language Design and Implementation*, June 2008. DOI: 10.1145/1375581.1375591. 83

[15] H. W. Cain and M. H. Lipasti. Memory ordering: A value-based approach. In *Proc. of the 31st Annual International Symposium on Computer Architecture*, June 2004. DOI: 10.1109/isca.2004.1310766. 58

[16] L. Ceze, J. Tuck, P. Montesinos, and J. Torrellas. BulkSC: Bulk enforcement of sequential consistency. In *Proc. of the 34th Annual International Symposium on Computer Architecture*, June 2007. DOI: 10.1145/1250662.1250697. 68

[17] S. Chakraborty and V. Vafeiadis. Grounding thin-air reads with event structures. In *POPL*, 2019. DOI: 10.1145/3290383. 86

[18] M. Dubois, C. Scheurich, and F. A. Briggs. Memory access buffering in multiprocessors. In *Proc. of the 13th Annual International Symposium on Computer Architecture*, pp. 434–42, June 1986. DOI: 10.1145/285930.285991. 81

[19] S. Flur, S. Sarkar, C. Pulte, K. Nienhuis, L. Maranget, K. E. Gray, A. Sezgin, M. Batty, and P. Sewell. Mixed-size concurrency: ARM, POWER, C/C++11, and SC. In *Proc. of the 44th ACM SIGPLAN Symposium on Principles of Programming Languages*, pp. 429–442, January 2017. DOI: 10.1145/3009837.3009839. 81

[20] K. Gharachorloo, D. Lenoski, J. Laudon, P. Gibbons, A. Gupta, and J. Hennessy. Memory consistency and event ordering in scalable shared-memory. In *Proc. of the 17th Annual International Symposium on Computer Architecture*, pp. 15–26, May 1990. DOI: 10.1109/isca.1990.134503. 72, 81

[21] C. Gniady, B. Falsafi, and T. Vijaykumar. Is SC + ILP = RC? In *Proc. of the 26th Annual International Symposium on Computer Architecture*, pp. 162–71, May 1999. DOI: 10.1109/isca.1999.765948. 58

[22] M. D. Hill. Multiprocessors should support simple memory consistency models. *IEEE Computer*, 31(8):28–34, August 1998. DOI: 10.1109/2.707614. 66

[23] IBM. Power ISA Version 2.06 Revision B. http://www.power.org/resources/downloads/ PowerISA_V2.06B_V2_PUBLIC.pdf, July 2010. 78, 81

[24] IBM Corporation. *Book E: Enhanced PowerPC Architecture, Version 0.91*, July 21, 2001. 78

[25] B. W. Kernighan and D. M. Ritchie. *The C Programming Language*, 2nd ed., Prentice Hall, 1988. 83

[26] S. Mador-Haim, R. Alur, and M. M. K. Martin. Generating litmus tests for contrasting memory consistency models. In *Proc. of the 22nd International Conference on Computer Aided Verification*, July 2010. DOI: 10.1007/978-3-642-14295-6_26. 82

[27] J. Manson, W. Pugh, and S. V. Adve. The Java memory model. In *Proc. of the 32nd Symposium on Principles of Programming Languages*, January 2005. http://dx.doi.org/10.1145/1040305.1040336.1040336 DOI: 10.1145/1040305.1040336. 83

[28] P. E. McKenney. *Is Parallel Programming Hard, And, If So, What Can You Do About It?* http://kernel.org/pub/linux/kernel/people/paulmck/perfbook/perfbook.2011.01.%05a.pd, 2011. 81

[29] A. Meixner and D. J. Sorin. Dynamic verification of memory consistency in cache-coherent multithreaded computer architectures. In *Proc. of the International Conference on Dependable Systems and Networks*, pp. 73–82, June 2006. DOI: 10.1109/dsn.2006.29. 81

[30] W. Pugh. The Java memory model is fatally flawed. *Concurrency: Practice and Experience*, 12(1):1–11, 2000. DOI: 10.1002/1096-9128(200005)12:6%3C445::aid-cpe484%3E3.0.co;2-a. 83

[31] The RISC-V Instruction Set Manual, Volume I: Unprivileged ISA https: //github.com/riscv/riscv-isa-manual/releases/download/draft-20190521- 21c6a14/riscv-spec.pdf DOI: 10.21236/ada605735. 77, 78

[32] J. Ševčík, and D. Aspinall. On validity of program transformations in the Java memory model. In *ECOOP*, 2008. DOI: 10.1007/978-3-540-70592-5_3. 85

[33] R. L. Sites, Ed. *Alpha Architecture Reference Manual*. Digital Press, 1992. 58, 81

[34] D. L. Weaver and T. Germond, Eds. *SPARC Architecture Manual (Version 9)*. PTR Prentice Hall, 1994. 58, 81, 82

CHAPTER 6

Coherence Protocols

In this chapter, we return to the topic of cache coherence that we introduced in Chapter 2. We defined coherence in Chapter 2, in order to understand coherence's role in supporting consistency, but we did not delve into how specific coherence protocols work or how they are implemented. This chapter discusses coherence protocols in general, before we move on to specific classes of protocols in the next two chapters. We start in Section 6.1 by presenting the big picture of how coherence protocols work, and then show how to specify protocols in Section 6.2. We present one simple, concrete example of a coherence protocol in Section 6.3 and explore the protocol design space in Section 6.4.

6.1 THE BIG PICTURE

The goal of a coherence protocol is to maintain coherence by enforcing the invariants introduced in Section 2.3 and restated here.

1. *Single-Writer, Multiple-Read (SWMR) Invariant.* For any memory location A, at any given (logical) time, there exists only a single core that may write to A (and can also read it) or some number of cores that may only read A.

2. *Data-Value Invariant.* The value of the memory location at the start of an epoch is the same as the value of the memory location at the end of its last read-write epoch.

To implement these invariants, we associate with each storage structure—each cache and the LLC/memory—a finite state machine called a *coherence controller*. The collection of these coherence controllers constitutes a distributed system in which the controllers exchange messages with each other to ensure that, for each block, the SWMR and data value invariants are maintained at all times. The interactions between these finite state machines are specified by the coherence protocol.

Coherence controllers have several responsibilities. The coherence controller at a cache, which we refer to as a *cache controller*, is illustrated in Figure 6.1. The cache controller must service requests from two sources. On the "core side," the cache controller interfaces to the processor core. The controller accepts loads and stores from the core and returns load values to the core. A cache miss causes the controller to initiate a coherence *transaction* by issuing a coherence *request* (e.g., request for read-only permission) for the block containing the location accessed by the core. This coherence request is sent across the interconnection network to one or more coherence controllers. A transaction consists of a request and the other message(s) that are exchanged in order to satisfy the request (e.g., a data response message sent from another coherence controller to the requestor). The types of transactions and the messages that are sent as part of each transaction depend on the specific coherence protocol.

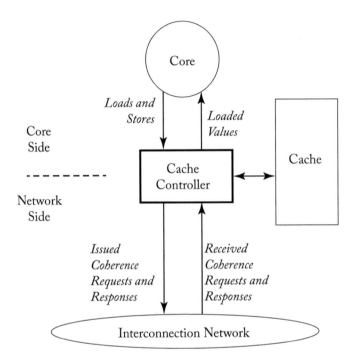

Figure 6.1: Cache controller.

On the cache controller's "network side," the cache controller interfaces to the rest of the system via the interconnection network. The controller receives coherence requests and coherence responses that it must process. As with the core side, the processing of incoming coherence messages depends on the specific coherence protocol.

The coherence controller at the LLC/memory, which we refer to as a *memory controller*, is illustrated in Figure 6.2. A memory controller is similar to a cache controller, except that it usually has only a network side. As such, it does not issue coherence requests (on behalf of loads or stores) or receive coherence responses. Other agents, such as I/O devices, may behave like cache controllers, memory controllers, or both depending upon their specific requirements.

Each coherence controller implements a set of finite state machines—logically one independent, but identical finite state machine per block—and receives and processes *events* (e.g., incoming coherence messages) depending upon the block's state. For an event of type E (e.g., a store request from the core to the cache controller) to block B, the coherence controller takes

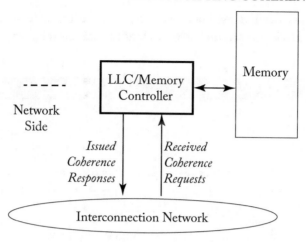

Figure 6.2: Memory controller.

actions (e.g., issues a coherence request for read-write permission) that are a function of E and of B's *state* (e.g., read-only). After taking these actions, the controller may change the state of B.

6.2   SPECIFYING COHERENCE PROTOCOLS

We specify a coherence protocol by specifying the coherence controllers. We could specify coherence controllers in any number of ways, but the particular behavior of a coherence controller lends itself to a tabular specification [9]. As shown in Table 6.1, we can specify a controller as a table in which rows correspond to block states and columns correspond to events. We refer to a state/event entry in the table as a *transition*, and a transition for event E pertaining to block B consists of (a) the actions taken when E occurs and (b) the next state of block B. We express transitions in the format "action/next state" and we may omit the "next state" portion if the next state is the current state. As an example of a transition in Table 6.1, if a store request for block B is received from the core and block B is in a read-only state (RO), then the table shows that the controller's transition will be to perform the action "issue coherence request for read-write permission (to block B)" and change the state of block B to RW.

The example in Table 6.1 is intentionally incomplete, for simplicity, but it illustrates the capability of a tabular specification methodology to capture the behavior of a coherence controller. To specify a coherence protocol, we simply need to completely specify the tables for the cache controllers and the memory controllers.

The differences between coherence protocols lie in the differences in the controller specifications. These differences include different sets of block states, transactions, events, and transitions. In Section 6.4, we describe the coherence protocol design space by exploring the options for each of these aspects, but we first specify one simple, concrete protocol.

Table 6.1: Tabular specification methodology. This is an incomplete specification of a cache coherence controller. Each entry in the table specifies the actions taken and the next state of the block.

		Events	
		Load request from core	Store request from core	Incoming coherence request to obtain block in read-write state
States	Not readable or writeable (N)	Issue coherence request for read-only permission/RO	Issue coherence requests for read-write permission/RW	\<No action\>
	Read-only (RO)	Give data from cache to core	Issue coherence request for read-write permission/RW	\<No action\>/N
	Read-write (RW)	Give data from cache to core	Write data to cache	Send block to requestor/N

6.3   EXAMPLE OF A SIMPLE COHERENCE PROTOCOL

To help understand coherence protocols, we now present a simple protocol. Our system model is the baseline system model from Section 2.1, but with the interconnection network restricted to being a shared bus: a shared set of wires on which a core can issue a message and have it observed by all cores and the LLC/memory.

Each cache block can be in one of two stable coherence states: I(nvalid) and V(alid). Each block at the LLC/memory can also be in one of two coherence states: I and V. At the LLC/memory, the state I denotes that all caches hold the block in state I, and the state V denotes that one cache holds the block in state V. There is also a single transient state for cache blocks, IV^D, discussed below. At system startup, all cache blocks and LLC/memory blocks are in state I. Each core can issue load and store requests to its cache controller; the cache controller will implicitly generate an Evict Block event when it needs to make room for another block. Loads and stores that miss in the cache initiate coherence transactions, as described below, to obtain a valid copy of the cache block. Like all the protocols in this primer, we assume a writeback cache; that is, when a store hits, it writes the store value only to the (local) cache and waits to write the entire block back to the LLC/memory in response to an Evict Block event.

There are two types of coherence transactions implemented using three types of bus messages: *Get* requests a block, *DataResp* transfers the block's data, and *Put* writes the block back to the memory controller. On a load or store miss, the cache controller initiates a Get transaction by sending a Get message and waiting for the corresponding DataResp message. The Get transaction is atomic in that no other transaction (either Get or Put) may use the bus between when the cache sends the Get and when the DataResp for that Get appears on the bus. On an Evict Block event, the cache controller sends a Put message, with the entire cache block, to the memory controller.

We illustrate the transitions between the stable coherence states in Figure 6.3. We use the prefaces "Own" and "Other" to distinguish messages for transactions initiated by the given cache controller vs. those initiated by other cache controllers. Note that if the given cache controller has the block in state V and another cache requests it with a Get message (denoted Other-Get), the owning cache must respond with a block (using a DataResp message, not shown) and transition to state I.

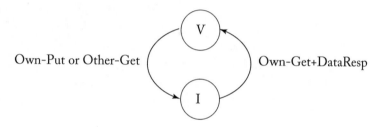

Figure 6.3: Transitions between stable states of blocks at cache controller.

Tables 6.2 and 6.3 specify the protocol in more detail. Shaded entries in the table denote impossible transitions. For example, a cache controller should never see its own Put request on the bus for a block that is in state V in its cache (as it should have already transitioned to state I).

The transient state IV^D corresponds to a block in state I that is waiting for data (via a DataResp message) before transitioning to state V. Transient states arise when transitions between stable states are not atomic. In this simple protocol, individual message sends and receives are atomic, but fetching a block from the memory controller requires sending a Get message and receiving a DataResp message, with an indeterminate gap in between. The IV^D state indicates that the protocol is waiting for a DataResp. We discuss transient states in more depth in Section 6.4.1.

This coherence protocol is simplistic and inefficient in many ways, but the goal in presenting this protocol is to gain an understanding of how protocols are specified. We use this specification methodology throughout this book when presenting different types of coherence protocols.

Table 6.2: Cache controller specification. Shaded entries are impossible and blank entries denote events that are ignored.

States	Core Events		Bus Events					
			Messages for Own Transactions			Messages for Other Cores' Transactions		
	Load or Store	Evict Block	Own-Get	DataResp for Own-Get	Own-Put	Other-Get	DataResp for Other-Get	Other-Put
I	Issue Get/ IV^D							
IV^D	Stall Load or Store	Stall Evict		Copy data into cache, perform Load or Store /V				
V	Perform Load or Store	Issue Put (with data) /I				Send DataResp /I		

Table 6.3: Memory controller specification

State	Bus Events	
	Get	Put
I	Send data block in DataResp message to requestor/V	
V		Update data block in memory/I

6.4   OVERVIEW OF COHERENCE PROTOCOL DESIGN SPACE

As mentioned in Section 6.1, a designer of a coherence protocol must choose the states, transactions, events, and transitions for each type of coherence controller in the system. The choice of stable states is largely independent of the rest of the coherence protocol. For example, there are two different classes of coherence protocols called snooping and directory, and an architect can design a snooping protocol or a directory protocol with the same set of stable states. We discuss stable states, independent of protocols, in Section 6.4.1. Similarly, the choice of transactions is also largely independent of the specific protocol, and we discuss transactions in Section 6.4.2. However, unlike the choices of stable states and transactions, the events, transitions and specific transient states are highly dependent on the coherence protocol and cannot be discussed in isolation. Thus, in Section 6.4.3, we discuss a few of the major design decisions in coherence protocols.

6.4.1   STATES

In a system with only one actor (e.g., a single core processor without coherent DMA), the state of a cache block is either valid or invalid. There might be two possible valid states for a cache block if there is a need to distinguish blocks that are *dirty*. A dirty block has a value that has been written more recently than other copies of this block. For example, in a two-level cache hierarchy with a write-back L1 cache, the block in the L1 may be dirty with respect to the stale copy in the L2 cache.

A system with multiple actors can also use just these two or three states, as in Section 6.3, but we often want to distinguish between different kinds of valid states. There are four characteristics of a cache block that we wish to encode in its state: validity, dirtiness, exclusivity, and ownership [10]. The latter two characteristics are unique to systems with multiple actors.

- Validity: A *valid* block has the most up-to-date value for this block. The block may be read, but it may only be written if it is also exclusive.

- Dirtiness: As in a single core processor, a cache block is *dirty* if its value is the most up-to-date value, this value differs from the value in the LLC/memory, and the cache controller is responsible for eventually updating the LLC/memory with this new value. The term *clean* is often used as the opposite of dirty.

- Exclusivity: A cache block is *exclusive*[1] if it is the only privately cached copy of that block in the system (i.e., the block is not cached anywhere else except perhaps in the shared LLC).

- Ownership: A cache controller (or memory controller) is the *owner* of a block if it is responsible for responding to coherence requests for that block. In most protocols, there is exactly one owner of a given block at all times. A block that is owned may not be evicted from a cache to make room for another block—due to a capacity or conflict miss—without giving the ownership of the block to another coherence controller. Non-owned blocks may be evicted silently (i.e., without sending any messages) in some protocols.

In this section, we first discuss some commonly used *stable states*—states of blocks that are not currently in the midst of a coherence transaction—and then discuss the use of *transient states* for describing blocks that are currently in the midst of transactions.

---

[1]The terminology here can be confusing, because there is a cache coherence state that is called "Exclusive," but there are other cache coherence states that are exclusive in the sense defined here.

Stable States

Many coherence protocols use a subset of the classic five state MOESI model first introduced by Sweazey and Smith [10]. These MOESI (often pronounced either "MO-sey" or "mo-EE-see") states refer to the states of blocks in a cache, and the most fundamental three states are MSI; the O and E states may be used, but they are not as basic. Each of these states has a different combination of the characteristics described previously.

- M(odified): The block is valid, exclusive, owned, and potentially dirty. The block may be read or written. The cache has the only valid copy of the block, the cache must respond to requests for the block, and the copy of the block at the LLC/memory is potentially stale.

- S(hared): The block is valid but not exclusive, not dirty, and not owned. The cache has a read-only copy of the block. Other caches may have valid, read-only copies of the block.

- I(nvalid): The block is invalid. The cache either does not contain the block or it contains a potentially stale copy that it may not read or write. In this primer, we do not distinguish between these two situations, although sometimes the former situation may be denoted as the "Not Present" state.

The most basic protocols use only the MSI states, but there are reasons to add the O and E states to optimize certain situations. We will discuss these optimizations in later chapters when we discuss snooping and directory protocols with and without these states. For now, here is the complete list of MOESI states:

- M(odified)

- O(wned): The block is valid, owned, and potentially dirty, but not exclusive. The cache has a read-only copy of the block and must respond to requests for the block. Other caches may have a read-only copy of the block, but they are not owners. The copy of the block in the LLC/memory is potentially stale.

- E(xclusive): The block is valid, exclusive, and clean. The cache has a read-only copy of the block. No other caches have a valid copy of the block, and the copy of the block in the LLC/memory is up-to-date. In this primer, we consider the block to be owned when it is in the Exclusive state, although there are protocols in which the Exclusive state is not treated as an ownership state. When we present MESI snooping and directory protocols in later chapters, we discuss the issues involved with making Exclusive blocks owners or not.

- S(hared)

- I(nvalid)

We illustrate a Venn diagram of the MOESI states in Figure 6.4. The Venn diagram shows which states share which characteristics. All states besides I are valid. M, O, and E are ownership states. Both M and E denote exclusivity, in that no other caches have a valid copy of the block. Both M and O indicate that the block is potentially dirty. Returning to the simplistic example in Section 6.3, we observe that the protocol effectively condensed the MOES states into the V state.

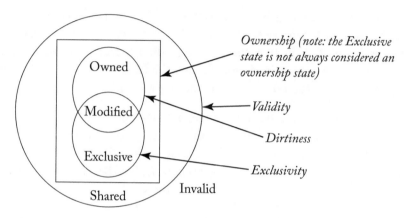

Figure 6.4: MOESI states.

The MOESI states, although quite common, are not an exhaustive set of stable states. For example, the F(orward) state is similar to the O state except that it is clean (i.e., the copy in the LLC/memory is up-to-date). There are many possible coherence states, but we focus our attention in this primer on the well-known MOESI states.

Transient States

Thus far, we have discussed only the stable states that occur when there is no current coherence activity for the block, and it is only these stable states that are used when referring to a protocol (e.g., "a system with a MESI protocol"). However, as we saw even in the example in Section 6.3, there may exist transient states that occur during the transition from one stable state to another stable state. In Section 6.3, we had the transient state IV^D (in I, going to V, waiting for DataResp). In more sophisticated protocols, we are likely to encounter dozens of transient states. We encode these states using a notation XY^Z, which denotes that the block is transitioning from stable state X to stable state Y, and the transition will not complete until an event of type Z occurs. For example, in a protocol in a later chapter, we use IM^D to denote that a block was previously in I and will become M once a D(ata) message arrives for that block.

States of Blocks in the LLC/Memory

The states that we have discussed thus far—both stable and transient—pertain to blocks residing in caches. Blocks in the LLC and memory also have states associated with them, and there are two general approaches to naming states of blocks in the LLC and memory. The choice of naming convention does not affect functionality or performance; it is simply a specification issue that can confuse an architect unfamiliar with the convention.

- Cache-centric: In this approach, which we believe to be the most common, the state of a block in the LLC and memory is an aggregation of the states of this block in the caches. For example, if a block is in all caches in I, then the LLC/memory state for this block is I. If a block is in one or more caches in S, then the LLC/memory state is S. If a block is in a single cache in M, then the LLC/memory state is M.

- Memory-centric: In this approach, the state of a block in the LLC/memory corresponds to the *memory controller's* permissions to this block (rather than the permissions of the caches). For example, if a block is in all caches in I, then the LLC/memory state for this block is O (not I, as in the cache-centric approach), because the LLC/memory behaves like an owner of the block. If a block is in one or more caches in S, then the LLC/memory state is also O, for the same reason. However, if the block is in a single cache in M or O, then the LLC/ memory state is I, because the LLC/memory has an invalid copy of the block.

All protocols in this primer use *cache-centric* names for the states of blocks in the LLC and memory.

Maintaining Block State

The system implementation must maintain the states associated with blocks in caches, the LLC, and memory. For caches and the LLC, this generally requires extending the per-block cache state by at most a few bits, since the number of stable states is generally small (e.g., 5 states for a MOESI protocol requires 3 bits per block). Coherence protocols may have many more transient states, but need maintain these states only for those blocks that have pending coherence transactions. Implementations typically maintain these transient states by adding additional bits to the miss status handling registers (MSHRs) or similar structures that are used to track these pending transactions [4].

For memory, it might appear that the much greater aggregate capacity would pose a significant challenge. However, many current multicore systems maintain an inclusive LLC, which means that the LLC maintains a copy of every block that is cached anywhere in the system (even "exclusive" blocks). With an inclusive LLC, memory does not need to explicitly represent the coherence state. If a block resides in the LLC, its state in memory is the same as its state in the LLC. If the block is not in the LLC, its state in memory is implicitly Invalid, because absence from an inclusive LLC implies that the block is not in any cache. The sidebar discusses how memory state was maintained in the days before multicores with inclusive LLCs. The above discussion of memory assumes a system with a single multicore chip, as does most of this primer. Systems with multiple multicore chips may benefit from explicit coherence state logically at memory.

6.4.2   TRANSACTIONS

Most protocols have a similar set of transactions, because the basic goals of the coherence controllers are similar. For example, virtually all protocols have a transaction for obtaining Shared (read-only) access to a block. In Table 6.4 we list a set of common transactions and, for each transaction, we describe the goal of the requestor that initiates the transaction. These transactions are all initiated by cache controllers that are responding to requests from their associated cores. In Table 6.5, we list the requests that a core can make to its cache controller and how these core requests can lead the cache controller into initiating coherence transactions.

> ### Sidebar: Before Multicores: Maintaining Coherence State at Memory
>
> Traditional, pre-multicore protocols needed to maintain coherence state for each block of memory, and they could not use the LLC as explained in Section 6.4.1. We briefly discuss several ways of maintaining this state and the associated engineering tradeoffs.
>
> **Augment Each Block of Memory with State Bits.** The most general implementation is to add extra bits to each block of memory to maintain the coherence state. If there are N possible states at memory, then each block needs $\log_2 N$ extra bits. Although this design is fully general and conceptually straightforward, it has several drawbacks. First, the extra bits may increase cost in two ways. Adding two or three extra bits is difficult with modern block-oriented DRAM chips, which are typically at least 4-bits wide and frequently much wider. Plus any change in the memory precludes using commodity DRAM modules (e.g., DIMMs), which significantly increases cost. Fortunately, for protocols that require only a few bits of state per block it is possible to store these using a modified ECC code. By maintaining ECC on a larger granularity (e.g., 512 bits rather than 64 bits), it is possible to free up enough code space to "hide" a handful of extra bits while using commodity DRAM modules [1, 5, 7]. The second drawback is that storing the state bits in DRAM means that obtaining the state incurs the full DRAM latency, even in the case that the most recent version of the block is stored in some other cache. In some cases, this may increase the latency of cache-to-cache coherence transfers. Finally, storing the state in DRAM means that all state changes require a DRAM read-modify-write cycle, which may impact both power and DRAM bandwidth.
>
> **Add Single State Bit per Block at Memory.** A design option used by the Synapse [3] was to distinguish the two stable states (I and V) using a single bit that is associated with

every block of memory. Few blocks are ever in transient states, and those states can be maintained with a small dedicated structure. This design is a subset of the more complete first design, with minimal storage cost.

Zero-bit logical OR. To avoid having to modify memory, we can have the caches reconstruct the memory state on-demand. The memory state of a block is a function of the block's state in every cache, so, if all of the caches aggregate their state, they can determine the memory state. The system can infer whether the memory is the owner of a block by having all of the cores send an "IsOwned?"[a] signal to a logical OR gate (or tree of OR gates) with a number of inputs equal to the number of caches. If the output of this OR is high, it denotes that a cache is owner; if the output is low, then memory is the owner. This solution avoids the need for any state to be maintained in memory. However, implementing a fast OR, either with logic gates or a wired-OR, can be difficult.

[a]This IsOwned signal is not to be confused with the Owned cache state. The IsOwned signal is asserted by a cache in a state of ownership, which includes the Owned, Modified, and Exclusive cache states.

Table 6.4: **Common transactions**

Transaction	Goal of Requestor
GetShared (GetS)	Obtain block in Shared (read-only) state
GetModified (GetM)	Obtain block in Modified (read-only) state
Upgrade (Upg)	Upgrade block state from read-only (Shared or Owned) to read-write (Modified); Upg (unlike GetM) does not require data to be sent to requestor
PutShared (PutS)	Evict block in Shared state[a]
PutExclusive (PutE)	Evict block in Exclusive state[a]
PutOwned (PutO)	Evict block in Owned state
PutModified (PutM)	Evict block in Modified state
[a] Some protocols do not require a coherence transaction to evict a Shared block and/or an Exclusive block (i.e., the PutS and/or PutE are "silent").	

Although most protocols use a similar set of transactions, they differ quite a bit in how the coherence controllers interact to perform the transactions. As we will see in the next section, in some protocols (e.g., snooping protocols) a cache controller initiates a GetS transaction by broadcasting a GetS request to all coherence controllers in the system, and whichever controller is currently the owner of the block responds to the requestor with a message that contains the desired data. Conversely, in other protocols (e.g., directory protocols) a cache controller initi-

Table 6.5: Common core requests to cache controller

Event	Response of (Typical) Cache Controller
Load	If cache hit, respond with data from cache; else initiate GetS transaction
Store	If cache hit in state E or M, write data into cache; else initiate GetM or Upg transaction
Atomic read-modify-write	If cache hit in state E or M, automically execute RMW semantics; else GetM or Upg transaction
Instruction fetch	If cache hit (in I-cache), respond with instruction from cache; else initiate GetS transaction
Read-only prefetch	If cache hit, ignore; else may optionally initiate GetS transaction[a]
Read-Write prefetch	If cache hit in state M, ignore; else may optionally initiate GetM or Upg transaction[a]
Replacement	Depending on state of block, initiate PutS, PutE, PutO, or PutM transaction

[a] A cache controller may choose to ignore a prefetch request from the core.

ates a GetS transaction by sending a unicast GetS message to a specific, pre-defined coherence controller that may either respond directly or may forward the request to another coherence controller that will respond to the requestor.

6.4.3   MAJOR PROTOCOL DESIGN OPTIONS

There are many different ways to design a coherence protocol. Even for the same set of states and transactions, there are many different possible protocols. The design of the protocol determines what events and transitions are possible at each coherence controller; unlike with states and transactions, there is no way to present a list of possible events or transitions that is independent from the protocol.

Despite the enormous design space for coherence protocols, there are two primary design decisions that have a major impact on the rest of the protocol, and we discuss them next.

Snooping vs. Directory

There are two main classes of coherence protocols: snooping and directory. We present a brief overview of these protocols now and defer in-depth coverage of them until Chapters 7 and 8, respectively.

- Snooping protocol: A cache controller initiates a request for a block by broadcasting a request message to all other coherence controllers. The coherence controllers collectively "do the right thing," e.g., sending data in response to another core's request if they are the owner. Snooping protocols rely on the interconnection network to deliver the broadcast messages in a consistent order to all cores. Most snooping protocols assume that requests arrive in a total order, e.g., via a shared-wire bus, but more advanced interconnection networks and relaxed orders are possible.

- Directory protocol: A cache controller initiates a request for a block by unicasting it to the memory controller that is the *home* for that block. The memory controller maintains a directory that holds state about each block in the LLC/memory, such as the identity of the current owner or the identities of current sharers. When a request for a block reaches the home, the memory controller looks up this block's directory state. For example, if the request is a GetS, the memory controller looks up the directory state to determine the owner. If the LLC/memory is the owner, the memory controller completes the transaction by sending a data response to the requestor. If a cache controller is the owner, the memory controller forwards the request to the owner cache; when the owner cache receives the forwarded request, it completes the transaction by sending a data response to the requestor.

The choice of snooping vs. directory involves making tradeoffs. Snooping protocols are logically simple, but they do not scale to large numbers of cores because broadcasting does not scale. Directory protocols are scalable because they unicast, but many transactions take more time because they require an extra message to be sent when the home is not the owner. In addition, the choice of protocol affects the interconnection network (e.g., classical snooping protocols require a total order for request messages).

Invalidate vs. Update
The other major design decision in a coherence protocol is to decide what to do when a core writes to a block. This decision is independent of whether the protocol is snooping or directory. There are two options.

- Invalidate protocol: When a core wishes to write to a block, it initiates a coherence transaction to invalidate the copies in all other caches. Once the copies are invalidated, the requestor can write to the block without the possibility of another core reading the block's old value. If another core wishes to read the block after its copy has been invalidated, it has to initiate a new coherence transaction to obtain the block, and it will obtain a copy from the core that wrote it, thus preserving coherence.

- Update protocol: When a core wishes to write a block, it initiates a coherence transaction to update the copies in all other caches to reflect the new value it wrote to the block.

Once again, there are tradeoffs involved in making this decision. Update protocols reduce the latency for a core to read a newly written block because the core does not need to initiate and wait for a GetS transaction to complete. However, update protocols typically consume substantially more bandwidth than invalidate protocols because update messages are larger than invalidate messages (an address and a new value, rather than just an address). Furthermore, update protocols greatly complicate the implementation of many memory consistency models. For example, preserving write atomicity (Section 5.5) becomes much more difficult when multiple caches must apply multiple updates to multiple copies of a block. Because of the complexity of update protocols, they are rarely implemented; in this primer, we focus on the far more common invalidate protocols.

Hybrid Designs

For both major design decisions, one option is to develop a hybrid. There are protocols that combine aspects of snooping and directory protocols [2, 6], and there are protocols that combine aspects of invalidate and update protocols [8]. The design space is rich and architects are not constrained to following any particular style of design.

6.5 REFERENCES

[1] A. Charlesworth. The Sun Fireplane SMP interconnect in the Sun 6800. In *Proc. of the 9th Hot Interconnects Symposium*, August 2001. DOI: 10.1109/his.2001.946691. 101

[2] P. Conway and B. Hughes. The AMD Opteron northbridge architecture. *IEEE Micro*, 27(2):10–21, March/April 2007. DOI: 10.1109/mm.2007.43. 105

[3] S. J. Frank. Tightly coupled multiprocessor system speeds memory-access times. *Electronics*, 57(1):164–169, January 1984. 101

[4] D. Kroft. Lockup-free instruction fetch/prefetch cache organization. In *Proc. of the 8th Annual Symposium on Computer Architecture*, May 1981. DOI: 10.1145/285930.285979. 100

[5] H. Q. Le et al. IBM POWER6 microarchitecture. *IBM Journal of Research and Development*, 51(6), 2007. DOI: 10.1147/rd.516.0639. 101

[6] M. M. K. Martin, D. J. Sorin, M. D. Hill, and D. A. Wood. Bandwidth adaptive snooping. In *Proc. of the 8th IEEE Symposium on High-Performance Computer Architecture*, pp. 251–262, January 2002. DOI: 10.1109/hpca.2002.995715. 105

[7] A. Nowatzyk, G. Aybay, M. Browne, E. Kelly, and M. Parkin. The S3.mp scalable shared memory multiprocessor. In *Proc. of the International Conference on Parallel Processing*, vol. I, pp. 1–10, August 1995. DOI: 10.1109/hicss.1994.323149. 101

[8] A. Raynaud, Z. Zhang, and J. Torrellas. Distance-adaptive update protocols for scalable shared-memory multiprocessors. In *Proc. of the 2nd IEEE Symposium on High-Performance Computer Architecture*, February 1996. DOI: 10.1109/hpca.1996.501197. 105

[9] D. J. Sorin, M. Plakal, M. D. Hill, A. E. Condon, M. M. Martin, and D. A. Wood. Specifying and verifying a broadcast and a multicast snooping cache coherence protocol. *IEEE Transactions on Parallel and Distributed Systems*, 13(6):556–578, June 2002. DOI: 10.1109/tpds.2002.1011412. 93

[10] P. Sweazey and A. J. Smith. A class of compatible cache consistency protocols and their support by the IEEE Futurebus. In *Proc. of the 13th Annual International Symposium on Computer Architecture*, pp. 414–423, June 1986. DOI: 10.1145/17356.17404. 97, 98

CHAPTER 7

Snooping Coherence Protocols

In this chapter, we present snooping coherence protocols. Snooping protocols were the first widely deployed class of protocols and they continue to be used in a variety of systems. Snooping protocols offer many attractive features, including low-latency coherence transactions and a conceptually simpler design than the alternative, directory protocols (Chapter 8).

We first introduce snooping protocols at a high level (Section 7.1). We then present a simple system with a complete but unsophisticated three-state (MSI) snooping protocol (Section 7.2). This system and protocol serve as a baseline upon which we later add system features and protocol optimizations. The protocol optimizations that we discuss include the additions of the Exclusive state (Section 7.3) and the Owned state (Section 7.4), as well as higher performance interconnection networks (Sections 7.5 and 7.6). We then discuss commercial systems with snooping protocols (Section 7.7) before concluding the chapter with a discussion of snooping and its future (Section 7.8).

Given that some readers may not wish to delve too deeply into snooping, we have organized the chapter such that readers may skim or skip Sections 7.3–7.6, if they so choose.

7.1 INTRODUCTION TO SNOOPING

Snooping protocols are based on one idea: all coherence controllers observe (*snoop*) coherence requests in the same order and collectively "do the right thing" to maintain coherence. By requiring that all requests to a given block arrive in order, a snooping system enables the distributed coherence controllers to correctly update the finite state machines that collectively represent a cache block's state.

Traditional snooping protocols broadcast requests to all coherence controllers, including the controller that initiated the request. The coherence requests typically travel on an ordered broadcast network, such as a bus. The ordered broadcast ensures that every coherence controller observes the same series of coherence requests in the same order, i.e., that there is a total order of coherence requests. Since a total order subsumes all per-block orders, this total order guarantees that all coherence controllers can correctly update a cache block's state.

To illustrate the importance of processing coherence requests in the same per-block order, consider the examples in Tables 7.1 and 7.2 where both core C1 and core C2 want to get the same block A in state M. In Table 7.1, all three coherence controllers observe the same per-block order of coherence requests and collectively maintain the single-writer–multiple-reader (SWMR) invariant. Ownership of the block progresses from the LLC/memory to core C1 to

Table 7.1: Snooping coherence example. All activity involves block A (denoted "A:").

Time	Core C1	Core C2	LLC/Memory
0	A:I	A:I	A:I (LLC/memory is owner)
1	A:GetM from Core C1/M	A:GetM from Core C1/I	A:GetM from Core C1/M (LLC/memory is not owner)
2	A:GetM from Core C2/I	A:GetM from Core C2/M	A:GetM from Core C2/M

Table 7.2: Snooping (In)coherence example. All activity involves block A (denoted "A:").

Time	Core C1	Core C2	LLC/Memory
0	A:I	A:I	A:I (LLC/memory is owner)
1	A:GetM from Core C1/M	**A:GetM from Core C2/M**	A:GetM from Core C1/M (LLC/memory is not owner)
2	A:GetM from Core C2/I	**A:GetM from Core C1/I**	A:GetM from Core C2/M

core C2. Every coherence controller independently arrives at the correct conclusion about the block's state as a result of each observed request. Conversely, Table 7.2 illustrates how incoherence might arise if core C2 observes a different per-block order of requests than core C1 and the LLC/memory. First, we have a situation in which both core C1 and core C2 are simultaneously in state M, which violates the SWMR invariant. Next, we have a situation in which no coherence controller believes it is the owner and thus a coherence request at this time would not receive a response (perhaps resulting in deadlock).

Traditional snooping protocols create a total order of coherence requests across *all blocks*, even though coherence requires only a per-block order of requests. Having a total order makes it easier to implement memory consistency models that require a total order of memory references, such as SC and TSO. Consider the example in Table 7.3 which involves two blocks A and B; each block is requested exactly once and so the system trivially observes per-block request orders. Yet because cores C1 and C2 observe the GetM and GetS requests out-of-order, this execution violates both the SC and TSO memory consistency models.

Table 7.3: Per-block order, coherence, and consistency. States and operations that pertain to address A are preceeded by the prefix "A:", and we denote a block A in state X with value V as "A:X[V]." If the value is stale, we omit it (e.g., "A:I"). Note that there are two blocks in this example, A and B, with A initially in state S at Core C2 and B initially in state M at Core C1.

Time	Core C1	Core C2	LLC/Memory
0	A:I B:M[0]	A:S[0] B:I	A:S[0] B:M
1	A:GetM from Core C1/M[0] **store A = 1** B:M[0]	A:S[0] B:I	A:S[0] B:M
2	A:M[1] **store B = 1** B:M[1]	A:S[0] B:I	A:GetM from Core C1/M B:M
3	A:M[1] B:GetS from Core C2/S[1]	A:S[0] B:I	A:M B:GetS from Core C2/S[1]
4	A:M[1] B:S[1]	A:S[0] B:GetS from Core C2/S[1] **r1 = B[1]**	A:M B:S[1]
5	A:M[1] B:S[1]	A:S[0] **r2 = A[0]** B:S[1]	A:M B:S[1]
6	A:M[1] B:S[1]	A:GetM from Core1/I B:S[1]	A:M B:S[1]
	r1 = 1, r2 = 0 violates SC and TSO		

Sidebar: How Snooping Depends on a Total Order of Coherence Requests

At first glance, the reader may assume that the problem in Table 7.3 arises because the SWMR invariant is violated for block A in cycle 1, since C1 has an M copy and C2 still has an S copy. However, Table 7.4 illustrates the same example, but enforces a total order of coherence requests. This example is identical until cycle 4, and thus has the same apparent SWMR violation. However, like the proverbial "tree in the forest," this violation does not cause a problem because it is not observed (i.e., there is "no one there to hear it"). Specifically, because the cores see both requests in the same order, C2 invalidates block A before it can see the new value for block B. Thus, when C2 reads block A, it must get the new value and therefore yields a correct SC and TSO execution.

Traditional snooping protocols use the total order of coherence requests to determine when, in a logical time based on snoop order, a particular request has been observed. In the example of Table 7.4, because of the total order, core C1 can infer that C2 will see the GetM for A before the GetS for B, and thus C2 does not need to send a specific acknowledgment message when it receives the coherence message. This implicit acknowledgment of request reception distinguishes snooping protocols from the directory protocols we study in the next chapter.

Table 7.4: Total order, coherence, and consistency. States and operations that pertain to address A are preceded by the prefix "A:", and we denote a block A in state X with value V as "A:X[V]." If the value is stale, we omit it (e.g., "A:I").

Time	Core C1	Core C2	LLC/Memory
0	A:I B:M[0]	A:S[0] B:I	A:S[0] B:M
1	A:GetM from Core C1/M[0] **store A = 1** B:M[0]	A:S[0] B:I	A:S[0] B:M
2	A:M[1] **store B = 1** B:M[1]	A:S[0] B:I	A:GetM from Core C1/M B:M
3	A:M[1] B:GetS from Core C2/S[1]	A:S[0] B:I	A:M B:GetS from Core C2/S[1]
4	A:M[1] B:S[1]	A:GetM from Core1/I B:I	A:M B:S[1]
5	A:M[1] B:S[1]	A:I B:GetS from Core C2/S[1] **r1 = B[1]**	A:M B:S[1]
6	A:GetS from Core C2/S[1] B:S[1]	A:GetS from Core C2/S[1] **r2 = A[1]** B:S[1]	A:GetS from Core C2/S[1] B:S[1]
	r1 = 1, r2 = 1 satisfies SC and TSO		

We discuss some of the subtler issues regarding the need for a total order in the sidebar.

Requiring that broadcast coherence requests be observed in a total order has important implications for the interconnection network used to implement traditional snooping protocols. Because many coherence controllers may simultaneously attempt to issue coherence requests, the interconnection network must serialize these requests into some total order. However the network determines this order, this mechanism becomes known as the protocol's *serialization (ordering) point*. In the general case, a coherence controller issues a coherence request, the network orders that request at the serialization point and broadcasts it to all controllers, and the issuing controller learns where its request has been ordered by snooping the stream of requests it receives from the controller. As a concrete and simple example, consider a system which uses a bus to broadcast coherence requests. Coherence controllers must use arbitration logic to ensure that only a single request is issued on the bus at once. This arbitration logic acts as the serialization point because it effectively determines the order in which requests appear on the bus. A subtle but important point is that a coherence request is ordered the instant the arbitration logic serializes it, but a controller may only be able to determine this order by snooping the bus to observe which other requests appear before and after its own request. Thus, coherence controllers may observe the total request order several cycles after the serialization point determines it.

Thus far, we have discussed only coherence requests, but not the responses to these requests. The reason for this seeming oversight is that the key aspects of snooping protocols revolve around the requests. There are few constraints on response messages. They can travel on a separate interconnection network that does not need to support broadcast nor have any ordering requirements. Because response messages carry data and are thus much longer than requests, there are significant benefits to being able to send them on a simpler, lower-cost network. Notably, response messages do not affect the serialization of coherence transactions. Logically, a coherence transaction—which consists of a broadcast request and a unicast response—occurs when the request is ordered, *regardless of when the response arrives at the requestor.* The time interval between when the request appears on the bus and when the response arrives at the requestor does affect the implementation of the protocol (e.g., during this gap, are other controllers allowed to request this block? If so, how does the requestor respond?), but it does not affect the serialization of the transaction.[1]

7.2   BASELINE SNOOPING PROTOCOL

In this section, we present a straightforward, unoptimized snooping protocol and describe its implementation on two different system models. The first, simple system model illustrates the basic approach for implementing snooping coherence protocols. The second, modestly more complex baseline system model illustrates how even relatively simple performance improvements may impact coherence protocol complexity. These examples provide insight into the key features of snooping protocols while revealing inefficiencies that motivate the features and optimizations presented in subsequent sections of this chapter. Sections 7.5 and 7.6 discuss how to adapt this baseline protocol for more advanced system models.

---

[1]This logical serialization of coherence transactions is analogous to the logical serialization of instruction execution in processor cores. Even when a core performs out-of-order execution, it still commits (serializes) instructions in program order.

7.2.1   HIGH-LEVEL PROTOCOL SPECIFICATION

The baseline protocol has only three stable states: M, S, and I. Such a protocol is typically referred to as an MSI protocol. Like the protocol in Section 6.3, this protocol assumes a write-back cache. A block is owned by the LLC/memory unless the block is in a cache in state M. Before presenting the detailed specification, we first illustrate a higher level abstraction of the protocol in order to understand its fundamental behaviors. In Figures 7.1 and 7.2, we show the transitions between the stable states at the cache and memory controllers, respectively.

There are three notational issues to be aware of. First, in Figure 7.1, the arcs are labeled with coherence requests that are observed on the bus. We intentionally omit other events, including loads, stores, and coherence responses. Second, the coherence events at the cache controller are labeled with either "Own" or "Other" to denote whether the cache controller observing the request is the requestor or not. Third, in Figure 7.2, we specify the state of a block at memory

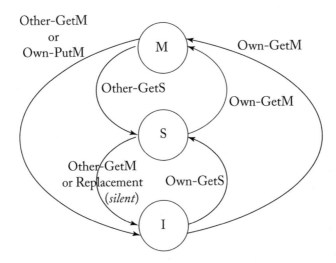

Figure 7.1: MSI: Transitions between stable states at cache controller.

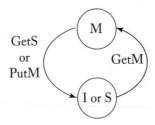

Figure 7.2: MSI: Transitions between stable states at memory controller.

using a cache-centric notation (e.g., a memory state of M denotes that there exists a cache with the block in state M).

7.2.2 SIMPLE SNOOPING SYSTEM MODEL: ATOMIC REQUESTS, ATOMIC TRANSACTIONS

Figure 7.3 illustrates the simple system model, which is nearly identical to the baseline system model introduced in Figure 2.1. The only difference is that the generic interconnection network from Figure 2.1 has been specified as a bus. Each core can issue load and store requests to its cache controller; the cache controller will choose a block to evict when it needs to make room for another block. The bus facilitates a total order of coherence requests that are snooped by all coherence controllers. Like the example in the previous chapter, this system model has atomicity properties that simplify the coherence protocol. Specifically, this system implements two atomicity properties which we define as *Atomic Requests* and *Atomic Transactions*. The *Atomic Requests* property states that a coherence request is ordered in the same cycle that it is issued. This property eliminates the possibility of a block's state changing—due to another core's coherence request—between when a request is issued and when it is ordered. The *Atomic Transactions*

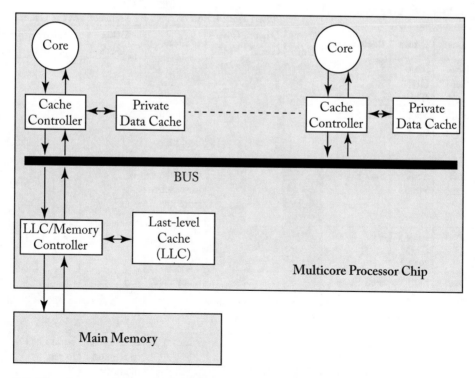

Figure 7.3: Simple snooping system mode.

property states that coherence transactions are *atomic* in that a subsequent request for the *same* block may not appear on the bus until after the first transaction completes (i.e., until after the response has appeared on the bus). Because coherence involves operations on a single block, whether or not the system permits subsequent requests to *different* blocks does not impact the protocol. Although simpler than most current systems, this system model resembles the SGI Challenge, a successful machine in the 1980s [5].

Detailed Protocol Specification

Tables 7.5 and 7.6 present the detailed coherence protocol for the simple system model. Compared to the high-level description in Section 7.2.1, the most significant difference is the addition of two transient states in the cache controller and one in the memory controller. This protocol has very few transient states because the atomicity constraints of the simple system model greatly limit the number of possible message interleavings.

Table 7.5: Simple snooping (atomic requests, atomic transactions): Cache controller

| States | Processor Core Events | | | Bus Events | | | | | | |
| | | | | Own Transaction | | | | Transactions for Other Cores | | |
	Load	Store	Replacement	Own-GetS	Own-GetM	Own-PutM	Data	Other-GetS	Other-GetM	Other-PutM
I	Issue GetS /IS^D	Issue GetM /IM^D								
IS^D	Stall Load	Stall Store	Stall Evict				Copy data into cache, load hit /S	(A)	(A)	(A)
IM^D	Stall Load	Stall Store	Stall Evict				Copy data into cache, store hit /M	(A)	(A)	(A)
S	Load hit	Issue GetM /SM^D	-/I						-/I	
SM^D	Load hit	Stall Store	Stall Evict				Copy data into cache, store hit /M	(A)	(A)	(A)
M	Load hit	Store Hit	Issue PutM, send Data to memory /I					Send Data to req and memory /S	Send Data to req /I	

Table 7.6: Simple snooping (atomic requests, atomic transactions): Memory controller

State	Bus Events			
	GetS	**GetM**	**PutM**	**Data from Owner**
IorS	Send data block in Data message to requestor/IorS	Send data block in Data message to requestor/M		
IorS^D	(A)	(A)		Update data block in memory/IorS
M	-/IorS^D		-/IorS^D	

Flashback to Quiz Question 6: In an MSI snooping protocol, a cache block may only be in one of three coherence states. *True or false?*

Answer: *False!* Even for the simplest system model, there are more than three states, because of transient states.

Shaded entries in the table denote impossible (or at least erroneous) transitions. For example, a cache controller should never receive a *Data* message for a block that it has not requested (i.e., a block in state I in its cache). Similarly, the *Atomic Transactions* constraint prevents another core from issuing a subsequent request before the current transaction completes; the table entries labeled "(A)" cannot occur due to this constraint. Blank entries denote legal transitions that require no action. These tables omit many implementation details that are not necessary for understanding the protocol. Also, in this protocol and the rest of the protocols in this chapter, we omit the event corresponding to Data for another core's transaction; a core never takes any action in response to observing Data on the bus for another core's transaction.

As with all MSI protocols, loads may be performed (i.e., hit) in states S and M, while stores hit only in state M. On load and store misses, the cache controller initiates coherence transactions by sending GetS and GetM requests, respectively.[2] The transient states IS^D, IM^D, and SM^D indicate that the request message has been sent, but the data response (Data) has not yet been received. In these transient states, because the requests have already been ordered, the transactions have already been ordered and the block is *logically* in state S, M, or M, respectively. A load or store must wait for the Data to arrive, though.[3] Once the data response appears on

---

[2]We do not include an Upgrade transaction in this protocol, which would optimize the S-to-M transition by not needlessly sending data to the requestor. Adding an Upgrade would be fairly straightforward for this system model with *Atomic Requests*, but it is significantly more complicated without *Atomic Requests*. We discuss this issue when we present a protocol without *Atomic Requests*.

[3]Technically, a store may be performed as soon as the request is ordered, so long as the newly stored value is not overwritten when the Data arrives. Similarly, a subsequent load to a newly written value is permitted.

the bus, the cache controller can copy the data block into the cache, transition to stable state S or M, as appropriate, and perform the pending load or store.

The system model's atomicity properties simplify cache miss handling in two ways. First, the *Atomic Requests* property ensures that when a cache controller seeks to upgrade permissions to a block—to go from I to S, I to M, or S to M—it can issue a request without worrying that another core's request might be ordered ahead of its own. Thus, the cache controller can transition immediately to state IS^D, IM^D, or SM^D, as appropriate, to wait for a data response. Similarly, the *Atomic Transactions* property ensures that no subsequent requests for a block will occur until after the current transaction completes, eliminating the need to handle requests from other cores while in one of these transient states.

A data response may come from either the memory controller or another cache that has the block in state M. A cache that has a block in state S can ignore GetS requests because the memory controller is required to respond, but must invalidate the block on GetM requests to enforce the coherence invariant. A cache that has a block in state M must respond to both GetS and GetM requests, sending a data response and transitioning to state S or state I, respectively.

The LLC/memory has two stable states, M and IorS, and one transient state $IorS^D$. In state IorS, the memory controller is the owner and responds to both GetS and GetM requests because this state indicates that no cache has the block in state M. In state M, the memory controller does not respond with data because the cache in state M is the owner and has the most recent copy of the data. However, a GetS in state M means that the cache controller will transition to state S, so the memory controller must also get the data, update memory, and begin responding to all future requests. It does this by transitioning immediately to the transient state $IorS^D$ and waits until it receives the data from the cache that owns it.

When the cache controller evicts a block due to a replacement decision, this leads to the protocol's two possible coherence downgrades: from S to I and from M to I. In this protocol, the S-to-I downgrade is performed "silently" in that the block is evicted from the cache without any communication with the other coherence controllers. In general, silent state transitions are possible only when all other coherence controllers' behavior remains unchanged; for example, a silent eviction of an owned block is not allowable. The M-to-I downgrade requires communication because the M copy of the block is the only valid copy in the system and cannot simply be discarded. Thus, another coherence controller (i.e., the memory controller) must change its state. To replace a block in state M, the cache controller issues a PutM request on the bus and then sends the data back to the memory controller. At the LLC, the block enters state $IorS^D$ when the PutM request arrives, then transitions to state IorS when the Data message arrives.[4] The *Atomic Requests* property simplifies the cache controller, by preventing an intervening request that might downgrade the state (e.g., another core's GetM request) before the PutM gets ordered on the bus. Similarly, the *Atomic Transactions* property simplifies the memory controller

---

[4]We make the simplifying assumption that these messages cannot arrive out of order at the memory controller.

by preventing other requests for the block until the PutM transaction completes and the memory controller is ready to respond to them.

Running Example

In this section, we present an example execution of the system to show how the coherence protocol behaves in a common scenario. We will use this example in subsequent sections both to understand the protocols and also to highlight differences between them. The example includes activity for just one block, and initially, the block is in state I in all caches and in state IorS at the LLC/memory.

In this example, illustrated in Table 7.7, cores C1 and C2 issue load and store instructions, respectively, that miss on the same block. Core C1 attempts to issue a GetS and core C2 attempts to issue a GetM. We assume that core C1's request happens to get serialized first and the *Atomic Transactions* property prevents core C2's request from reaching the bus until C1's request completes. The memory controller responds to C1 to complete the transaction on cycle 3. Then, core C2's GetM is serialized on the bus; C1 invalidates its copy and the memory controller responds to C2 to complete that transaction. Lastly, C1 issues another GetS. C2, the owner, responds with the data and changes its state to S. C2 also sends a copy of the data to the memory controller because the LLC/memory is now the owner and needs an up-to-date copy of the block. At the end of this execution, C1 and C2 are in state S and the LLC/memory is in state IorS.

7.2.3 BASELINE SNOOPING SYSTEM MODEL: NON-ATOMIC REQUESTS, ATOMIC TRANSACTIONS

The baseline snooping system model, which we use for most of the rest of this chapter, differs from the simple snooping system model by permitting non-atomic requests. Non-atomic requests arise from a number of implementation optimizations, but most commonly due to inserting a message queue (or even a single buffer) between the cache controller and the bus. By separating when a request is issued from when it is ordered, the protocol must address a window of vulnerability that did not exist in the simple snooping system. The baseline snooping system model preserves the *Atomic Transactions* property, which we do not relax until Section 7.5.

We present the detailed protocol specification, including all transient states, in Tables 7.8 and 7.9. Compared to the protocol for the simple snooping system in Section 7.2.2, the most significant difference is the much larger number of transient states. Relaxing the *Atomic Requests* property introduces numerous situations in which a cache controller observes a request from another controller on the bus in between issuing its coherence request and observing its own coherence request on the bus.

Taking the I-to-S transition as an example, the cache controller issues a GetS request and changes the block's state from I to IS^{AD}. Until the requesting cache controller's own GetS is observed on the bus and serialized, the block's state is *effectively* I. That is, the requestor's

Table 7.7: Simple snooping: Example execution. All activity is for one block.

Cycle	Core C1	Core C2	LLC/memory	Request on Bus	Data on Bus
Initial	I	I	IorS		
1	Load miss; issue GetS/IS^D				
2				GetS (C1)	
3		Store miss; stall due to *Atomic Transactions*	Send response to C1		
4					Data from LLC/ mem
5	Copy data to cache; perform load/S	Issue GetM/IM^D			
6				GetM (C2)	
7	-/I		Send response to C2/M		
8					Data from LLC/ mem
9		Copy data to cache; perform store/M			
10	Load miss: issue GetS/IS^D				
11				GetS (C1)	
12		Send data to C1 and to LLC/mem/S	-/IorS^D		
13					Data from C2
14	Copy data from C2; perform load/S		Copy data from C2/IorS		

block is treated as if it were in I; loads and stores cannot be performed and coherence requests from other nodes must be ignored. Once the requestor observes its own GetS, the request is ordered and block is logically S, but loads cannot be performed because the data has not yet arrived. The cache controller changes the block's state to IS^D and waits for the data response from the previous owner. Because of the *Atomic Transactions* property, the data message is the next coherence message (to the same block). Once the data response arrives, the transaction is complete and the requestor changes the block's state to the stable S state and performs the load. The I-to-M transition proceeds similarly to this I-to-S transition.

Table 7.8: MSI snooping protocol with atomic transactions-cache controller. A shaded entry labeled "(A)" denotes that this transition is impossible because transactions are atomic on bus.

	Load	Store	Replacement	Own-GetS	Own-GetM	Own-PutM	Other-GetS	Other-GetM	Other-PutM	Own Data Response
I	Issue GetS/IS^{AD}	Issue GetM/IM^{AD}					-	-	-	
IS^{AD}	Stall	Stall	Stall	-/IS^D			-	-	-	
IS^D	Stall	Stall	Stall				(A)	(A)		-/S
IM^{AD}	Stall	Stall	Stall		-/IM^D		-	-	-	
IM^D	Stall	Stall	Stall				(A)	(A)		-/M
S	Hit	Issue GetM/SM^{AD}	-/I				-	-/I	-	
SM^{AD}	Hit	Stall	Stall		-/SM^D		-	-/IM^{AD}	-	
SM^D	Hit	Stall	Stall				(A)	(A)		-/M
M	Hit	Hit	Issue PutM/MI^A				Send data to requestor and to memory/S	Send data to requestor/I	-	
MI^A	Hit	Hit	Stall			Send data to memory/I	Send data to requestor and to memory/II^A	Send data to requestor/II^A		
II^A	Stall	Stall	Stall			Send NoData to memory/I	-	-		

The transition from S to M illustrates the potential for state changes to occur during the window of vulnerability. If a core attempts to store to a block in state S, the cache controller issues a GetM request and transitions to state SM^{AD}. The block remains effectively in state S, so loads may continue to hit and the controller ignores GetS requests from other cores. However, if another core's GetM request gets ordered first, the cache controller must transition the state to IM^{AD} to prevent further load hits. The window of vulnerability during the S-to-M transition complicates the addition of an Upgrade transaction, as we discuss in the sidebar.

Table 7.9: MSI snooping protocol with atomic transactions-memory controller. A shaded entry labeled "(A)" denotes that this transition is impossible because transactions are atomic on bus.

	GetS	GetM	PutM	Data from Owner	NoData
IorS	Send data to requestor	Send data to requestor/M	-/IorS^D		
IorS^D	(A)	(A)		Write data to LLC/memory/IorS	-/IorS
M	-/IorS^D	-	-/M^D		
M^D	(A)	(A)		Write data to LLC/IorS	-/M

Sidebar: Upgrade Transactions in Systems Without Atomic Requests

For the protocol with *Atomic Requests*, an Upgrade transaction is an efficient way for a cache to transition from Shared to Modified. The Upgrade request invalidates all shared copies, and it is much faster than issuing a GetM, because the requestor needs to wait only until the Upgrade is serialized (i.e., the bus arbitration latency) rather than wait for data to arrive from the LLC/memory.

However, without *Atomic Requests*, adding an Upgrade transaction becomes more difficult because of the window of vulnerability between issuing a request and when the request is serialized. The requestor may lose its shared copy due to an Other-GetM or Other-Upgrade that is serialized during this window of vulnerability. The simplest solution to this problem is to change the block's state to a new state in which it waits for its own Upgrade to be serialized. When its Upgrade is serialized, which will invalidate other S copies (if any) but will not return data, the core must then issue a subsequent GetM request to transition to M.

Handling Upgrades more efficiently is difficult, because the LLC/memory needs to know when to send data. Consider the case in which cores C0 and C2 have a block A shared and both seek to upgrade it and, at the same time, core C1 seeks to read it. C0 and C2 issue Upgrade requests and C1 issues a GetS request. Suppose they serialize on the bus as C0, C1, and C2. C0's Upgrade succeeds, so the LLC/memory (in state IorS) should change its state to M but not send any data, and C2 should invalidate its S copy. C1's GetS finds the block in state M at C0, which responds with the new data value and updates the LLC/memory back to state IorS. C2's Upgrade finally appears, but because it has lost its shared copy, it needs the LLC/memory to respond. Unfortunately, the LLC/memory is in state IorS and cannot tell that this Upgrade needs data. Alternatives exist to solve this issue, but are outside the scope of this primer.

The window of vulnerability also affects the M-to-I coherence downgrade, in a much more significant way. To replace a block in state M, the cache controller issues a PutM request and changes the block state to MI^A; unlike the protocol in Section 7.2.2, it does not immediately send the data to the memory controller. Until the PutM is observed on the bus, the block's state is effectively M and the cache controller must respond to other cores' coherence requests for the block. In the case where no intervening coherence requests arrive, the cache controller responds to observing its own PutM by sending the data to the memory controller and changing the block state to state I. If an intervening GetS or GetM request arrives before the PutM is ordered, the cache controller must respond as if it were in state M and then transition to state II^A to wait for its PutM to appear on the bus. Once it sees its PutM, intuitively, the cache controller should simply transition to state I because it has already given up ownership of the block. Unfortunately, doing so will leave the memory controller stuck in a transient state because it also receives the PutM request. Nor can the cache controller simply send the data anyway because doing so might overwrite valid data.[5] The solution is for the cache controller to send a special NoData message to the memory controller when it sees its PutM while in state II^A. The NoData message indicates to the memory controller that it is coming from a non-owner and lets the memory controller exit its transient state. The memory controller is further complicated by needing to know which stable state it should return to if it receives a NoData message. We solve this problem by adding a second transient memory state M^D. Note that these transient states represent an exception to our usual transient state naming convention. In this case, state X^D indicates that the memory controller should revert to state X when it receives a NoData message (and move to state IorS if it receives a data message).

7.2.4   RUNNING EXAMPLE

Returning to the running example, illustrated in Table 7.10, core C1 issues a GetS and core C2 issues a GetM. Unlike the previous example (in Table 7.7), eliminating the *Atomic Requests* property means that both cores issue their requests and change their state. We assume that core C1's request happens to get serialized first, and the *Atomic Transactions* property ensures that C2's request does not appear on the bus until C1's transaction completes. After the LLC/memory responds to complete C1's transaction, core C2's GetM is serialized on the bus. C1 invalidates its copy and the LLC/memory responds to C2 to complete that transaction. Lastly, C1 issues another GetS. When this GetS reaches the bus, C2, the owner, responds with the data and changes its state to S. C2 also sends a copy of the data to the memory controller because the LLC/memory is now the owner and needs an up-to-date copy of the block. At the end of this execution, C1 and C2 are in state S and the LLC/memory is in state IorS.

---

[5]Consider the case in which core C1 has a block in M and issues a PutM, but core C2 does a GetM and core C3 does a GetS, both of which are ordered before C1's PutM. C2 gets the block in M, modifies the block, and then in response to C3's GetS, updates the LLC/memory with the updated block. When C1's PutM is finally ordered, writing the data back would overwrite C2's update.

Table 7.10: Baseline snooping: Example execution

Cycle	Core C1	Core C2	LLC/memory	Request on Bus	Data on Bus
1	Issue GetS/IS^{AD}				
2		Issue GetM/IM^{AD}			
3				GetS (C1)	
4	-/IS^D		Send data to C1/IorS		
5					Data from LLC/mem
6	Copy data from LLC/mem/S			GetM (C2)	
7	-/I	-/IM^D	Send data to C2/M		
8					Data from LLC/mem
9		Copy data from LLC/mem/M			
10	Issue GetS/IS^{AD}				
11				GetS (C1)	
12	-/IS^D	Send data to C1 and to LLC/mem/S	-/IorS^D		
13					Data from C2
14	Copy data from C2/S		Copy data from C2/IorS		

7.2.5 PROTOCOL SIMPLIFICATIONS

This protocol is relatively straightforward and sacrifices performance to achieve this simplicity. The most significant simplification is the use of atomic transactions on the bus. Having atomic transactions eliminates many possible transitions, denoted by "(A)" in the tables. For example, when a core has a cache block in state IM^D, it is not possible for that core to observe a coherence request for that block from another core. If transactions were not atomic, such events could occur and would force us to redesign the protocol to handle them, as we show in Section 7.5.

Another notable simplification that sacrifices performance involves the event of a store request to a cache block in state S. In this protocol, the cache controller issues a GetM and changes the block state to SM^{AD}. A higher performance but more complex solution would use an upgrade transaction, as discussed in the earlier sidebar.

7.3   ADDING THE EXCLUSIVE STATE

There are many important protocol optimizations, which we discuss in the next several sections. More casual readers may want to skip or skim these sections on first reading. One very commonly used optimization is to add the Exclusive (E) state, and in this section, we describe how to create a MESI snooping protocol by augmenting the baseline protocol from Section 7.2.3 with the E state. Recall from Chapter 6 that if a cache has a block in the Exclusive state, then the block is valid, read-only, clean, exclusive (not cached elsewhere), and owned. A cache controller may silently change a cache block's state from E to M without issuing a coherence request.

7.3.1   MOTIVATION

The Exclusive state is used in almost all commercial coherence protocols because it optimizes a common case. Compared to an MSI protocol, a MESI protocol offers an important advantage in the situation in which a core first reads a block and then subsequently writes it. This is a typical sequence of events in many important applications, including single-threaded applications. In an MSI protocol, on a load miss, the cache controller will initiate a GetS transaction to obtain read permission; on the subsequent store, it will then initiate a GetM transaction to obtain write permission. However, a MESI protocol enables the cache controller to obtain the block in state E, instead of S, in the case that the GetS occurs when no other cache has access to the block. Thus, a subsequent store does not require the GetM transaction; the cache controller can silently upgrade the block's state from E to M and allow the core to write to the block. The E state can thus eliminate half of the coherence transactions in this common scenario.

7.3.2   GETTING TO THE EXCLUSIVE STATE

Before explaining how the protocol works, we must first figure out how the issuer of a GetS determines that there are no other sharers and thus that it is safe to go directly to state E instead of state S. There are at least two possible solutions:

- Adding a wired-OR "sharer" signal to bus: when the GetS is ordered on the bus, all cache controllers that share the block assert the "sharer" signal. If the requestor of the GetS observes that the "sharer" signal is asserted, the requestor changes its block state to S; else, the requestor changes its block state to E. The drawback to this solution is having to implement the wired-OR signal. This additional shared wire might not be problematic in this baseline snooping system model that already has a shared wire bus, but it would greatly complicate implementations that do not use shared wire buses (Section 7.6).

- Maintaining extra state at the LLC: an alternative solution is for the LLC to distinguish between states I (no sharers) and S (one or more sharers), which was not needed for the MSI protocols. In state I, the memory controller responds with data that is specially labeled as being Exclusive; in state S, the memory controller responds with data that is unlabeled. However, maintaining the S state exactly is challenging, since the LLC must detect when the last sharer relinquishes its copy. First, this requires that a cache controller issues a PutS message when it evicts a block in state S. Second, the memory controller must maintain a count of the sharers as part of the state for that block. This is much more complex and bandwidth intensive than our previous protocols, which allowed for silent evictions of blocks in S. A simpler, but less complete, alternative allows the LLC to conservatively track sharers; that is, the memory controller's state S means that there are zero-or-more caches in state S. The cache controller silently replaces blocks in state S, and thus the LLC stays in S even after the last sharer has been replaced. If a block in state M is written back (with a PutM), the state of the LLC block becomes I. This "conservative S" solution forgoes some opportunities to use the E state (i.e., when the last sharer replaces its copy before another core issues a GetM), but it avoids the need for explicit PutS transactions and still captures many important sharing patterns.

In the MESI protocol we present in this section, we choose the most implementable option—maintaining a conservative S state at the LLC—to both avoid the engineering problems associated with implementing wired-OR signals in high-speed buses and avoid explicit PutS transactions.

7.3.3   HIGH-LEVEL SPECIFICATION OF PROTOCOL

In Figures 7.4 and 7.5, we show the transitions between stable states in the MESI protocol. The MESI protocol differs from the baseline MSI protocol at both the cache and LLC/memory. At the cache, a GetS request transitions to S or E, depending upon the state at the LLC/memory when the GetS is ordered. Then, from state E, the block can be silently changed to M. In this protocol, we use a PutM to evict a block in E, instead of using a separate PutE; this decision helps keep the protocol specification concise, and it has no impact on the protocol functionality.

The LLC/memory has one more stable state than in the MSI protocol. The LLC/memory must now distinguish between blocks that are shared by zero or more caches (the conservative S state) and those that are not shared at all (I), instead of merging those into one single state as was done in the MSI protocol.

In this primer, we consider the E state to be an ownership state, which has a significant effect on the protocol. There are, however, protocols that do not consider the E state to be an ownership state, and the sidebar discusses the issues involved in such protocols.

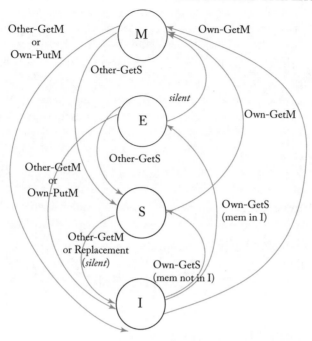

Figure 7.4: **MESI:** Transitions between stable states at cache controller.

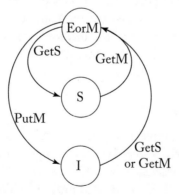

Figure 7.5: **MESI:** Transitions between stable states at memory controller.

Sidebar: MESI Snooping if E is Non-ownership State

If the E state is not considered an ownership state (i.e., a block in E is owned by the LLC/memory), then the protocol must figure out which coherence controller should respond to a request after the memory controller has given a block to a cache in state E. Because the transition from state E to state M is silent, the memory controller cannot know whether the cache holds the block in E, in which case the LLC/memory is the owner, or in M, in which case the cache is the owner. If a GetS or GetM is serialized on the bus at this point, the cache can easily determine whether it is the owner and should respond, but the memory controller cannot make this same determination.

One solution to this problem is to have the LLC/memory wait for the cache to respond. When a GetS or GetM is serialized on the bus, a cache with the block in state M responds with data. The memory controller waits a fixed amount of time and, if no response appears in that window of time, the memory controller deduces that it is the owner and that it must respond. If a response from a cache does appear, the memory controller does not respond to the coherence request. This solution has a couple drawbacks, including potentially increased latency for responses from memory. Some implementations hide some or all of this latency by speculatively prefetching the block from memory, at the expense of increased memory bandwidth, power, and energy. A more significant drawback is having to design the system such that the caches' response latency is predictable and short.

7.3.4 DETAILED SPECIFICATION

In Tables 7.11 and 7.12, we present the detailed specification of the MESI protocol, including transient states. Differences with respect to the MSI protocol are highlighted with boldface font. The protocol adds to the set of cache states just the stable E state and the transient state EI^A, but there are several more LLC/memory states, including an extra transient state.

This MESI protocol shares all of the same simplifications present in the baseline MSI protocol. Coherence transactions are still atomic, etc.

7.3.5 RUNNING EXAMPLE

We now return to the running example, illustrated in Table 7.13. The execution differs from the MSI protocol almost immediately. When C1's GetS appears on the bus, the LLC/memory is in state I and can thus send C1 Exclusive data. C1 observes the Exclusive data on the bus and changes its state to E (instead of S, as in the MSI protocol). The rest of the execution proceeds similarly to the MSI example, with minor transient state differences.

7.4 ADDING THE OWNED STATE

A second important optimization is the Owned state, and in this section, we describe how to create a MOSI snooping protocol by augmenting the baseline protocol from Section 7.2.3 with the O state. Recall from Chapter 6 that if a cache has a block in the Owned state, then the block is valid, read-only, dirty, and the cache is the owner, i.e., the cache must respond to

Table 7.11: MESI Snooping protocol—cache controller. A shaded entry labeled "(A)" denotes that this transition is impossible because transactions are atomic on bus.

	Load	Store	Replacement	OwnGetS	OwnGetM	Own-PutM	Other-GetS	Other-GetM	Other-PutM	Own Data Response	Own Data Response (exclusive)
I	Issue GetS/IS^{AD}	Issue GetM/IM^{AD}					-	-	-		
IS^{AD}	Stall	Stall	Stall	-/IS^D			-	-	-		
IS^D	Stall	Stall	Stall				(A)	(A)	(A)	-/S	-/E
IM^{AD}	Stall	Stall	Stall		-/IM^D		-	-	-		
IM^D	Stall	Stall	Stall				(A)	(A)	(A)	-/M	
S	Hit	Issue GetM/SM^{AD}	-/I				-	-/I	-		
SM^{AD}	Hit	Stall	Stall		-/SM^D		-	-/IM^{AD}	-		
SM^D	Hit	Stall	Stall				(A)	(A)		-/M	
E	**Hit**	**Hit/M**	**Issue PutM/EI^A**				**Send data to requestor and to memory/S**	**Send data to requestor/I**	-		
M	Hit	Hit	Issue PutM/MI^A				Send data to requestor and to memory/S	Send data to requestor/I	-		
MI^A	Hit	Hit	Stall			Send data to memory/I	Send data to requestor and to memory/II^A	Send data to requestor/II^A	-		
EI^A	Hit	Stall	Stall			Send No-Data-E to memory/I	**Send data to requestor and to memory/II^A**	Send data to requestor/II^A	-		
II^A	Stall	Stall	Stall			Send No-Data to memory/I	-	-	-		

Table 7.12: MESI Snooping protocol—memory controller. A shaded entry labeled "(A)" denotes that this transition is impossible because transactions are atomic on bus.

	GetS	GetM	PutM	Data	NoData	NoData-E
I	Send data to requestor /EorM	Send data to requestor /EorM	$-/I^D$			
S	Send data to requestor	Send data to requestor /EorM	$-/S^D$			
EorM	$-/S^D$	-	$-/EorM^D$			
I^D	(A)	(A)	(A)	Write data to memory/I	-/I	-/I
S^D	(A)	(A)	(A)	Write data to memory/S	-/S	-/S
$EorM^D$	(A)	(A)	(A)	Write data to memory/I	-/EorM	-/I

coherence requests for the block. We maintain the same system model as the baseline snooping MSI protocol; transactions are atomic but requests are not atomic.

7.4.1   MOTIVATION

Compared to an MSI or MESI protocol, adding the O state is advantageous in one specific and important situation: when a cache has a block in state M or E and receives a GetS from another core. In the MSI protocol of Section 7.2.3 and the MESI protocol of Section 7.3, the cache must change the block state from M or E to S and send the data to *both* the requestor *and* the memory controller. The data must be sent to the memory controller because the responding cache relinquishes ownership (by downgrading to state S) and the LLC/memory becomes the owner and thus must thus have an up-to-date copy of the data with which to respond to subsequent requests.

Adding the O state achieves two benefits: (1) it eliminates the extra data message to update the LLC/memory when a cache receives a GetS request in the M (and E) state, and (2) it eliminates the potentially unnecessary write to the LLC (if the block is written again before being written back to the LLC). Historically, for multi-chip multiprocessors, there was a third benefit, which was that the O state allows subsequent requests to be satisfied by the cache instead of by the far-slower memory. Today, in a multicore with an inclusive LLC, as in the system model in this primer, the access latency of the LLC is not nearly as long as that of off-chip DRAM memory. Thus, having a cache respond instead of the LLC is not as big of a benefit as having a cache respond instead of memory.

We now present a MOSI protocol and show how it achieves these two benefits.

Table 7.13: MESI: Example execution

Cycle	Core C1	Core C2	LLC/memory	Request on Bus	Data on Bus
1	Issue GetS/IS^{AD}				
2		Issue GetM/IM^{AD}			
3				GetS (C1)	
4	-/IS^D		Send exclusive data to C1/EorM		
5					Exclusive data from LLC/mem
6	**Copy data from LLC/mem /E**			**GetM (C2)**	
7	Send data to C2/I	-/M^D	-/EorM		
8					Data from C1
9		Copy data from C1/M			
10	Issue GetS/IS^{AD}				
11				GetS (C1)	
12	-/IS^D	Send data to C1 and to LLC/mem /S	-/S^D		
13					Data from C2
14	Copy data from C2/S		**Copy data from C2/S**		

7.4.2 HIGH-LEVEL PROTOCOL SPECIFICATION

We specify a high-level view of the transitions between stable states in Figures 7.6 and 7.7. The key difference is what happens when a cache with a block in state M receives a GetS from another core. In a MOSI protocol, the cache changes the block state to O (instead of S) and retains ownership of the block (instead of transferring ownership to the LLC/memory). Thus, the O state enables the cache to avoid updating the LLC/memory.

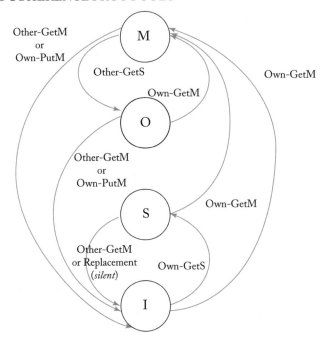

Figure 7.6: **MOSI**: Transitions between stable states at cache controller.

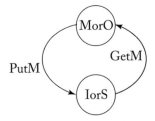

Figure 7.7: **MOSI**: Transitions between stable states at memory controller.

7.4.3   DETAILED PROTOCOL SPECIFICATION

In Tables 7.14 and 7.15, we present the detailed specification of the MOSI protocol, including transient states. Differences with respect to the MSI protocol are highlighted with boldface font. The protocol adds two transient cache states in addition to the stable O state. The transient OI^A state helps handle replacements of blocks in state O, and the transient OM^A state handles upgrades back to state M after a store. The memory controller has no additional transient states, but we rename what had been the M state to MorO because the memory controller does not need to distinguish between these two states.

Table 7.14: MOSI Snooping protocol—cache controller. A shaded entry labeled "(A)" denotes that this transition is impossible because transactions are atomic on bus.

	Load	Store	Replacement	Own-GetS	Own-GetM	Own-PutM	Other-GetS	Other-GetM	Other-PutM	Own Data Response
I	Issue GetS /IS^{AD}	Issue GetM /IM^{AD}					-	-	-	
IS^{AD}	Stall	Stall	Stall	-/IS^D			-	-	-	
IS^D	Stall	Stall	Stall				(A)	(A)	(A)	-/S
IM^{AD}	Stall	Stall	Stall		-/IM^D		-	-	-	
IM^D	Stall	Stall	Stall				(A)	(A)	(A)	-/M
S	Hit	Issue GetM /SM^{AD}	-/I				-	-/I	-	
SM^{AD}	Hit	Stall	Stall		-/SM^D		-	-/IM^{AD}	-	
SM^D	Hit	Stall	Stall				(A)	(A)	(A)	-/M
O	Hit	Issue GetM /OM^A	Issue PutM /OI^A				Send data to requestor	Send data to requestor/I	-	
OM^A	Hit	Stall	Stall		-/M		Send data to requestor	Send data to requestor /IM^{AD}	-	
M	Hit	Hit	IssuePutM /MI^A				Send data to requestor/O	Send data to requestor/I	-	
MI^A	Hit	Hit	Stall			Send data to memory /I	Send data to requestor /OI^A	Send data to requestor /II^A	-	
OI^A	Hit	Stall	Stall			Send data to memory /I	Send data to requestor	Send data to requestor /II^A	-	
II^A	Stall	Stall	Stall			Send No-Data to memory/I	-	-	-	

To keep the specification as concise as possible, we consolidate the PutM and PutO transactions into a single PutM transaction. That is, a cache evicts a block in state O with a PutM. This decision has no impact on the protocol's functionality, but does help to keep the tabular specification readable.

This MOSI protocol shares all of the same simplifications present in the baseline MSI protocol. Coherence transactions are still atomic, etc.

Table 7.15: MOSI Snooping protocol—memory controller. A shaded entry labeled "(A)" denotes that this transition is impossible because transactions are atomic on bus.

	GetS	GetM	PutM	Data from Owner	NoData
IorS	Send data to requestor	**Send data to requestor/MorO**	-/IorS^D		
IorS^D	(A)	(A)		Write data to memory/IorS	-/IorS
MorO	-	-	-/MorO^D		
MorO^D	(A)	(A)		**Write data to memory/IorS**	-/MorO

7.4.4   RUNNING EXAMPLE

In Table 7.16, we return to the running example that we introduced for the MSI protocol. The example proceeds identically to the MSI example until C1's second GetS appears on the bus. In the MOSI protocol, this second GetS causes C2 to respond to C1 and change its state to O (instead of S). C2 retains ownership of the block and does not need to copy the data back to the LLC/ memory (unless and until it evicts the block, not shown).

7.5   NON-ATOMIC BUS

The baseline MSI protocol, as well as the MESI and MOSI variants, all rely on the *Atomic Transactions* assumption. This atomicity greatly simplifies the design of the protocol, but it sacrifices performance.

7.5.1   MOTIVATION

The simplest way to implement atomic transactions is to use a shared-wire bus with an atomic bus protocol; that is, all bus transactions consist of an indivisible request-response pair. Having an atomic bus is analogous to having an unpipelined processor core; there is no way to overlap activities that could proceed in parallel. Figure 7.8 illustrates the operation of an atomic bus. Because a coherence transaction occupies the bus until the response completes, an atomic bus trivially implements atomic transactions. However, the throughput of the bus is limited by the sum of the latencies for a request and response (including any wait cycles between request and response, not shown). Considering that a response could be provided by off-chip memory, this latency bottlenecks bus performance.

Figure 7.9 illustrates the operation of a pipelined, non-atomic bus. The key advantage is not having to wait for a response before a subsequent request can be serialized on the bus, and thus the bus can achieve much higher bandwidth using the same set of shared wires. How-

Table 7.16: MOSI: Example execution

Cycle	Core C1 (C1)	Core C2 (C2)	LLC/memory	Request on Bus	Data on Bus
1	Issue GetS/IS^{AD}				
2		Issue GetM/IM^{AD}			
3				GetS (C1)	
4	-/IS^D		Send data to C1/IorS		
5					Data from LLC/mem
6	Copy data from LLC/mem /S			GetM (C2)	
7	-/I	-/IM^D	**Send data to C2/MorO**		
8					Data from LLC/mem
9		Copy data from LLC/mem /M			
10	Issue GetS/IS^{AD}				
11				GetS (C1)	
12	-/IS^D	**Send data to C1/O**	-/MorO		
13					Data from C2
14	Copy data from C2/S				

ever, implementing atomic transactions becomes much more difficult (but not impossible). The atomic transactions property restricts concurrent transactions to the same block, but not different blocks. The SGI Challenge enforced atomic transactions on a pipelined bus using a fast table lookup to check whether or not another transaction was already pending for the same block.

7.5.2   IN-ORDER VS. OUT-OF-ORDER RESPONSES

One major design issue for a non-atomic bus is whether it is pipelined or split-transaction. A *pipelined* bus, as illustrated in Figure 7.9, provides responses in the same order as the requests. A

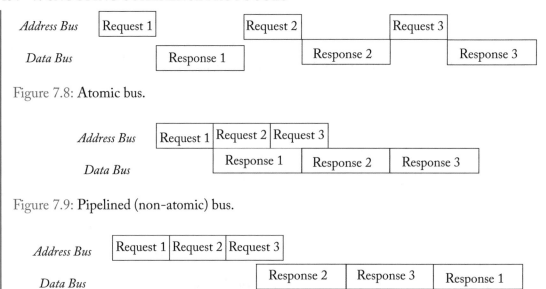

Figure 7.8: Atomic bus.

Figure 7.9: Pipelined (non-atomic) bus.

Figure 7.10: Split transaction (non-atomic) bus.

split-transaction bus, illustrated in Figure 7.10, can provide responses in an order different from the request order.

The advantage of a split-transaction bus, with respect to a pipelined bus, is that a low-latency response does not have to wait for a long-latency response to a prior request. For example, if Request 1 is for a block owned by memory and not present in the LLC and Request 2 is for a block owned by an on-chip cache, then forcing Response 2 to wait for Response 1, as a pipelined bus would require, incurs a performance penalty.

One issue raised by a split-transaction bus is matching responses with requests. With an atomic bus, it is obvious that a response corresponds to the most recent request. With a pipelined bus, the requestor must keep track of the number of outstanding requests to determine which message is the response to its request. With a split-transaction bus, the response must carry the identity of the request or the requestor.

7.5.3   NON-ATOMIC SYSTEM MODEL

We assume a system like the one illustrated in Figure 7.11. The request bus and the response bus are split and operate independently. Each coherence controller has connections to and from both buses, with the exception that the memory controller does not have a connection to make requests. We draw the FIFO queues for buffering incoming and outgoing messages because it is important to consider them in the coherence protocol. Notably, if a coherence controller stalls when processing an incoming request from the request bus, then all requests behind it

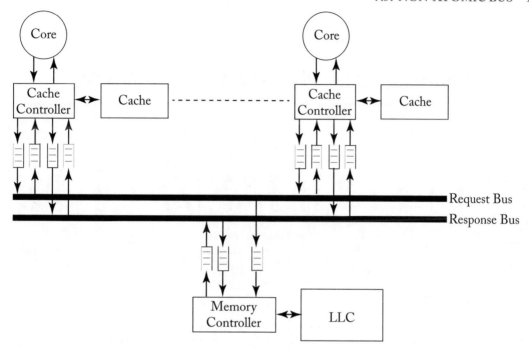

Figure 7.11: System model with split-transaction bus.

(serialized after the stalled request) will not be processed by that coherence controller until it processes the currently stalled request. These queues are processed in a strict FIFO fashion, regardless of message type or address.

7.5.4   AN MSI PROTOCOL WITH A SPLIT-TRANSACTION BUS

In this section, we modify the baseline MSI protocol for use in a system with a split-transaction bus. Having a split-transaction bus does not change the transitions between stable states, but it has a large impact on the detailed implementation. In particular, there are many more possible transitions.

In Tables 7.17 and 7.18, we specify the protocol. Several transitions are now possible that were not possible with the atomic bus. For example, a cache can now receive an Other-GetS for a block it has in state IS^D. All of these newly possible transitions are for blocks in transient states in which the cache is awaiting a data response; while waiting for the data, the cache first observes another coherence request for the block. Recall from Section 7.1 that a transaction is ordered based on when its request is ordered on the bus, not when the data arrives at the requestor. Thus, in each of these newly possible transitions, the cache has already effectively completed its transaction but just happens to not have the data yet. Returning to our example of IS^D, the

Table 7.17: **MSI Snooping protocol with split-transaction bus—cache controller**

	Load	Store	Replacement	Own-GetS or Own-GetM	Own-GetM	Own-PutM	Other-GetS	Other-GetM	Other-PutM	Own Data Response (for own request)
I	Issue GetS /IS^{AD}	Issue GetM /IM^{AD}					-	-	-	
IS^{AD}	Stall	Stall	Stall	-/IS^D			-	-	-	-/IS^A
IS^D	Stall	Stall	Stall				-	Stall		Load hit/S
IS^A	**Stall**	**Stall**	**Stall**	**Load hit/S**			-	-		
IM^{AD}	Stall	Stall	Stall		-/IM^D		-	-	-	-/IM^A
IM^D	Stall	Stall	Stall				Stall	Stall		Store hit/M
IM^A	**Stall**	**Stall**	**Stall**		**Store hit/M**		-	-		
S	Hit	Issue GetM /SM^{AD}	-/I				-	-/I		
SM^{AD}	Hit	Stall	Stall	-/SM^D			-	-/IM^{AD}		-/SM^A
SM^D	Hit	Stall	Stall				**Stall**	**Stall**		Store hit/M
SM^A	**Hit**	**Stall**	**Stall**	**Store hit/M**			-	-/IM^A		
M	Hit	Hit	Issue PutM /MI^A				Send data to requestor and to memory /S	Send data to requestor /I		
MI^A	Hit	Hit	Stall			Send data to requestor /I	Send data to requestor and to memory /II^A	Send data to requestor /II^A		
II^A	Stall	Stall	Stall			-/I	-	-	-	

Table 7.18: MSI Snooping protocol with split-transaction bus—memory controller

	GetS	GetM	PutM from Owner	PutM from Non-Owner	Data
IorS	Send data to requestor	Send data to requestor, set Owner to requestor/M		-	
M	Clear Owner /IorS^D	Set Owner to requestor	Clear Owner/IorS^D	-	**Write data to memory/IorS^A**
IorS^D	**Stall**	**Stall**	**Stall**	-	Write data to memory/IorS
IorS^A	**Clear Owner /IorS**	-	**Clear Owner/IorS**	-	

cache block is effectively in S. Thus, the arrival of an Other-GetS in this state requires no action to be taken because a cache with a block in S need not respond to an Other-GetS.

The newly possible transitions other than the above example, however, are more complicated. Consider a block in a cache in state IM^D when an Other-GetS is observed on the bus. The cache block is effectively in state M and the cache is thus the owner of the block but does not yet have the block's data. Because the cache is the owner, the cache must respond to the Other-GetS, yet the cache cannot respond until it receives the data. The simplest solution to this situation is for the cache to stall processing of the Other-GetS until the data response arrives for its Own-GetM. At that point, the cache block will change to state M and the cache will have valid data to send to the requestor of the Other-GetS.

For the other newly possible transitions, at both the cache controller and the memory controller, we also choose to stall until data arrives to satisfy the in-flight request. This is the simplest approach, but it raises three issues. First, it sacrifices some performance, as we discuss in the next section.

Second, stalling raises the potential of deadlock. If a controller can stall on a message while awaiting another event (message arrival), the architect must ensure that the awaited event will eventually occur. Circular chains of stalls can lead to deadlock and must be avoided. In our protocol in this section, controllers that stall are guaranteed to receive the messages that un-stall them. This guarantee is easy to see because the controller has already seen its own request, the stall only affects the request network, and the controller is waiting for a Data message on the response network.

The third issue raised by stalling coherence requests is that, perhaps surprisingly, it enables a requestor to observe a response to its request before processing its own request. Consider the example in Table 7.19. Core C1 issues a GetM for block X and changes the state of X to IM^{AD}. C1 observes its GetM on the bus and changes state to IM^D. The LLC/memory is the owner of X and takes a long time to retrieve the data from memory and put it on the bus. In the meanwhile, core C2 issues a GetM for X that gets serialized on the bus but cannot be processed by C1 (i.e., C1 stalls). C1 issues a GetM for block Y that then gets serialized on the bus. This GetM for Y is queued up behind the previously stalled coherence request at C1 (the GetM from C2) and thus C1 cannot process its own GetM for Y. However, the owner, C2, can process this GetM for Y and responds quickly to C1. Thus, C1 can observe the response to its GetM for Y before processing its request. This possibility requires the addition of transient states. In this example, core C1 changes the state of block Y from IM^{AD} to IM^A. Similarly, the protocol also needs to add transient states IS^A and SM^A. In these transient states, in which the response is observed before the request, the block is effectively in the prior state. For example, a block in IM^A is logically in state I because the GetM has not been processed yet; the cache controller does not respond to an observed GetS or GetM if the block is in IM^A. We contrast IM^A with IM^D—in IM^D, the block is logically in M and the cache controller must respond to observed GetS or GetM requests once data arrives.

This protocol has one other difference with respect to the previous protocols in this chapter, and the difference pertains to PutM transactions. The situation that is handled differently is when a core, say, core C1, issues a PutM, and a GetS or GetM from another core for the same block gets ordered before C1's PutM. C1 transitions from state MI^A to II^A before it observes its own PutM. In the atomic protocols earlier in this chapter, C1 observes its own PutM and sends a NoData message to the LLC/memory. The NoData message informs the LLC/memory that the PutM transaction is complete (i.e., it does not have to wait for data). C1 cannot send a Data message to the LLC/ memory in this situation because C1's data are stale and the protocol cannot send the LLC/memory stale data that would then overwrite the up-to-date value of the data. In the non-atomic protocols in this chapter, we augment the state of each block in the LLC with a field that holds the identity of the current owner of the block. The LLC updates the owner field of a block on every transaction that changes the block's ownership. Using the owner field, the LLC can identify situations in which a PutM from a non-owner is ordered on the bus; this is exactly the same situation in which C1 is in state II^A when it observes its PutM. Thus, the LLC knows what happened and C1 does not have to send a NoData message to the LLC. We chose to modify how PutM transactions are handled in the non-atomic protocols, compared to the atomic protocols, for simplicity. Allowing the LLC to directly identify this situation is simpler than requiring the use of NoData messages; with a non-atomic protocol, there can be a large number of NoData messages in the system and NoData messages can arrive before their associated PutM requests.

Table 7.19: Example: Response before request. Initially, block X is in state I in both caches and block Y is in state M at core C2.

Cycle	Core C1 (C1)	Core C2 (C2)	LLC/memory	Request on Bus	Data on Bus
Initial	X:I Y:I	X:I Y:M	X:I Y:M		
1	X: store miss; issue GetM/IM^{AD}				
2				X: GetM (C1)	
3	X: process GetM (C1) / IM^D	X: process GetM (C1) - ignore	X: process GetM (C1) - LLC miss, start accessing X from DRAM		
4		X: store miss; issue GetM/ IM^{AD}			
5	Y: store miss; issue GetM/ IM^{AD}			X: GetM (C2)	
6	X: stall on GetM (C2)	X: process GetM (C2) / IM^D	X: process GetM (C2) - ignore	Y: GetM (C1)	
7	Y: queue GetM (C1)	Y: process GetM (C1) - send data to C1/I	Y: process GetM (C1) - ignore		
8					Y: data from C2
9	**Y: write data into cache/ IM^A**				
10			X: LLC miss completes, send data to C1		
11					X: data from LLC
12	X: write data into cache/M; Perform store				
13	X: (unstall) process GetM (C2) - send data to C2/I				
14	**Y: process (in-order) GetM (C1)/M ; perform store**				X: data from C1
15		X: write data into cache/M; perform store			

7.5.5 AN OPTIMIZED, NON-STALLING MSI PROTOCOL WITH A SPLIT-TRANSACTION BUS

As mentioned in the previous section, we sacrificed some performance by stalling on the newly possible transitions of the system with the split-transaction bus. For example, a cache with a block in state IS^D stalled instead of processing an Other-GetM for that block. However, it is possible that there are one or more requests after the Other-GetM, to other blocks, that the cache could process without stalling. By stalling a request, the protocol stalls all requests after the stalled request and delays those transactions from completing. Ideally, we would like a coherence controller to process requests behind a request that is stalled, but recall that—to support a total order of memory requests—snooping requires coherence controllers to observe and process requests in the order received. Reordering is not allowed.

The solution to this problem is to process all messages, in order, instead of stalling. Our approach is to add transient states that reflect messages that the coherence controller has received but must remember to complete at a later event. Returning to the example of a cache block in IS^D, if the cache controller observes an Other-GetM on the bus, then it changes the block state to IS^DI (which denotes "in I, going to S, waiting for data, and when data arrives will go to I"). Similarly, a block in IM^D that receives an Other-GetS changes state to IM^DS and must remember the requestor of the Other-GetS. When the data arrive in response to the cache's GetM, the cache controller sends the data to the requestor of the Other-GetS and changes the block's state to S.

In addition to the proliferation of transient states, a non-stalling protocol introduces a potential livelock problem. Consider a cache with a block in IM^DS that receives the data in response to its GetM. If the cache *immediately* changes the block state to S and sends the data to the requestor of the Other-GetS, it does not get to perform the store for which it originally issued its GetM. If the core then re-issues the GetM, the same situation could arise again and again, and the store might never perform. To guarantee that this livelock cannot arise, we require that a cache in IS^DI, IM^DI, IM^DS, or IM^DSI (or any comparable state in a protocol with additional stable coherence states) perform one load or store to the block when it receives the data for its request.[6] After performing one load or store, it may then change state and forward the block to another cache. We defer a more in-depth treatment of livelock to Section 9.3.2.

We present the detailed specification of the non-stalling MSI protocol in Tables 7.20 and 7.21. The most obvious difference is the number of transient states. There is nothing inherently complicated about any of these states, but they do add to the overall complexity of the protocol.

We have not removed the stalls from the memory controller because it is not feasible. Consider a block in $IorS^D$. The memory controller observes a GetM from core C1 and currently stalls.

---

[6]The load or store must be performed *if and only if* that load or store was the oldest load or store in program order when the coherence request was first issued. We discuss this issue in more detail in Section 9.3.2.

Table 7.20: Optimized MSI snooping with split-transaction bus—cache controller

	Load	Store	Replacement	Own-GetS	Own-GetM	Own-PutM	Other-GetS	Other-GetM	Other-PutM	Own Data Response
I	Issue GetS /IS^{AD}	Issue GetM /IM^{AD}					-	-	-	
IS^{AD}	Stall	Stall	Stall	-/IS^D			-	-	-	-/IS^A
IS^D	Stall	Stall	Stall				-	-/IS^{D}I		Load hit/S
IS^A	**Stall**	**Stall**	**Stall**	**Load Hit/S**			-	.		
IS^{D}I	**Stall**	**Stall**	**Stall**				-	.		Load hit/I
IM^{AD}	Stall	Stall	Stall		-/IM^D		-	.	-	-/IM^A
IM^D	Stall	Stall	Stall				-/IM^{D}S	-/IM^{D}I		Store hit/M
IM^A	**Stall**	**Stall**	**Stall**		**Store hit/M**		-	-	-	
IM^{D}I	**Stall**	**Stall**	**Stall**				-	-		**Store hit, send data to GetM requestor/I**
IM^{D}S	**Stall**	**Stall**	**Stall**				-	-/IM^{D}SI		**Store hit, send data to GetS requestor and mem/S**
IM^{D}SI	**Stall**	**Stall**	**Stall**				-	-		**Store hit, send data to GetS requestor and mem/I**
S	Hit	Issue GetM/ SM^{AD}	-/I				-	-/I		
SM^{AD}	Hit	Stall	Stall	-/SM^D			-	-/IM^{AD}		-/SM^A
SM^D	Hit	Stall	Stall				-/SM^{D}S	-/SM^{D}I		Store hit/M
SM^A	**Hit**	**Stall**	**Stall**		**Store hit/M**		-	-/IM^A		
SM^{D}I	**Hit**	**Stall**	**Stall**				-	-		**Store hit, send data to GetM requestor/I**
SM^{D}S	**Hit**	**Stall**	**Stall**				-	-/SM^{D}SI		**Store hit, send data to GetS requestor and mem/S**
SM^{D}SI	**Hit**	**Stall**	**Stall**				-	-		**Store hit, send data to GetS requestor and mem/I**
M	Hit	Hit	Issue PutM/ MI^A				Send data to requestor and to memory/S	Send data to requestor/I		
MI^A	Hit	Hit	Stall			Send data to requestor /I	Send data to requestor and to memory /II^A	Send data to requestor/II^A		
II^A	Stall	Stall	Stall			-/I	-	-	-	

Table 7.21: Optimized MSI snooping with split-transaction bus—memory controller

	GetS	GetM	PutM from Owner	PutM from Non-Owner	Data
IorS	Send data to requestor	Send data to requestor, set Owner to requestor/M		-	
M	Clear Owner /IorS^D	Set Owner to requestor	Clear Owner /IorS^D	-	**Write data to memory/IorS^A**
IorS^D	**Stall**	**Stall**	**Stall**	-	Write data to memory/IorS
IorS^A	**Clear Owner /IorS**	-	**Clear Owner /IorS**	-	

However, it would appear that we could simply change the block's state to IorS^{D}M while waiting for the data. Yet, while in IorS^{D}M, the memory controller could observe a GetS from core C2. If the memory controller does not stall on this GetS, it must change the block state to IorS^{D}MIorS^D. In this state, the memory controller could observe a GetM from core C3. There is no elegant way to bound the number of transient states needed at the LLC/memory to a small number (i.e., smaller than the number of cores) and so, for simplicity, we have the memory controller stall.

7.6   OPTIMIZATIONS TO THE BUS INTERCONNECTION NETWORK

So far in this chapter we have assumed system models in which there exists a single shared-wire bus for coherence requests and responses or dedicated shared-wire buses for requests and responses. In this section, we explore two other possible system models that enable improved performance.

7.6.1   SEPARATE NON-BUS NETWORK FOR DATA RESPONSES

We have emphasized the need of snooping systems to provide a total order of broadcast coherence requests. The example in Table 7.2 showed how the lack of a total order of coherence requests can lead to incoherence. However, there is no such need to order coherence *responses*, nor is there a need to broadcast them. Thus, coherence responses could travel on a separate network that does not support broadcast or ordering. Such networks include crossbars, meshes, tori, butterflies, etc.

There are several advantages to using a separate, non-bus network for coherence responses.

- Implementability: it is difficult to implement high-speed shared-wire buses, particularly for systems with many controllers on the bus. Other topologies can use point-to-point links.

- Throughput: a bus can provide only one response at a time. Other topologies can have multiple responses in-flight at a time.

- Latency: using a bus for coherence responses requires that each response incur the latency to arbitrate for the bus. Other topologies can allow responses to be sent immediately without arbitration.

7.6.2   LOGICAL BUS FOR COHERENCE REQUESTS

Snooping systems require that there exist a total order of broadcast coherence requests. A shared-wire bus for coherence requests is the most straightforward way to achieve this total order of broadcasts, but it is not the only way to do so. There are two ways to achieve the same totally ordered broadcast properties as a bus (i.e., a logical bus) without having a physical bus.

- Other topologies with physical total order: a shared-wire bus is the most obvious topology for achieving a total order of broadcasts, but other topologies exist. One notable example is a tree with the coherence controllers at the leaves of the tree. If all coherence requests are unicasted to the root of the tree and then broadcast down the tree, then each coherence controller observes the same total order of coherence broadcasts. The serialization point in this topology is the root of the tree. Sun Microsystems used a tree topology in its Starfire multiprocessor [3], which we discuss in detail in Section 7.7.

- Logical total order: a total order of broadcasts can be obtained even without a network topology that naturally provides such an order. The key is to order the requests in logical time. Martin et al. [6] designed a snooping protocol, called Timestamp Snooping, that can function on any network topology. To issue a coherence request, a cache controller broadcasts it to every coherence controller and labels the broadcast with the logical time at which the broadcast message should be ordered. The protocol must ensure that (a) every broadcast has a distinct logical time, (b) coherence controllers process requests in logical time order (even when they arrive out of this order in physical time), and (c) no request at logical time T can arrive at a controller after that controller has passed logical time T. Agarwal et al. proposed a similar scheme called In-Network Snoop Ordering (INSO) [1].

Flashback to Quiz Question 7: A snooping cache coherence protocol requires the cores to communicate on a bus. *True or false?*
Answer: *False!* Snooping requires a totally ordered broadcast network, but that functionality can be implemented without a physical bus.

7.7   CASE STUDIES

We present two examples of real-world snooping systems: the Sun Starfire E10000 and the IBM Power5.

7.7.1   SUN STARFIRE E10000

Sun Microsystems's Starfire E10000 [3] is an interesting example of a commercial system with a snooping protocol. The coherence protocol itself is not that remarkable; the protocol is a typical MOESI snooping protocol with write-back caches. What distinguishes the E10000 is how it was designed to scale up to 64 processors. The architects innovated based on three important observations, which we discuss in turn.

First, shared-wire snooping buses do not scale to large numbers of cores, largely due to electrical engineering constraints. In response to this observation, the E10000 uses only point-to-point links instead of buses. Instead of broadcasting coherence requests on physical (shared-wire) buses, the E10000 broadcasts coherence requests on a *logical* bus. The key insight behind snooping protocols is that they require a total order of coherence requests, but this total order does not require a physical bus. As illustrated in Figure 7.12, the E10000 implements a logical bus as a tree, in which the processors are the leaves. All links in the tree are point-to-point, thus eliminating the need for buses. A processor unicasts a request up to the top of the tree, where it is serialized and then broadcast down the tree. Because of the serialization at the root, the tree provides totally ordered broadcast. A given request may arrive at two processors at different times, which is fine; the important constraint is that the processors observe the same total order of requests.

The second observation made by the E10000 architects is that greater coherence request bandwidth can be achieved by using multiple (logical) buses, while still maintaining a total order of coherence requests. The E10000 has four logical buses, and coherence requests are address-interleaved across them. A total order is enforced by requiring processors to snoop the logical buses in a fixed, pre-determined order.

Third, the architects observed that data response messages, which are much larger than request messages, do not require the totally ordered broadcast network required for coherence requests.

Many prior snooping systems implemented a data bus, which needlessly provides both broadcasting and total ordering, while limiting bandwidth. To improve bandwidth, the E10000 implements the data network as a crossbar. Once again, there are point-to-point links instead of buses, and the bandwidth of the crossbar far exceeds what would be possible with a bus (physical or logical).

The architecture of the E10000 has been optimized for scalability, and this optimized design requires the architects to reason about non-atomic requests and non-atomic transactions.

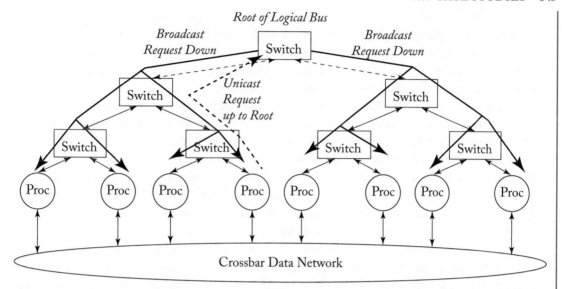

Figure 7.12: Starfire E10000 (drawn with only eight processors for clarity). A coherence request is unicast up to the root, where it is serialized, before being broadcast down to all processors.

7.7.2   IBM POWER5

The IBM Power5 [8] is a 2-core chip in which both cores share an L2 cache. Each Power5 chip has a fabric bus controller (FBC) that enables multiple Power5 chips to be connected together to create larger systems. Large systems contain up to eight nodes, where each node is a multi-chip module (MCM) with four Power5 chips.

Viewed abstractly, the IBM Power5 appears to use a fairly typical MESI snooping protocol implemented atop a split-transaction bus. However, this simplistic description misses several unique features that are worth discussing. In particular, we focus on two aspects: the ring topology of the interconnection network and the addition of novel variants of the MESI coherence states.

Snooping Coherence on a Ring

The Power5 uses an interconnection network that is quite different from what we have discussed thus far, and these differences have important impacts on the coherence protocol. Most significantly, the Power5 connects nodes with three unidirectional rings, which are used for carrying three types of messages: requests, snoop responses/decision messages, and data. Unidirectional rings do not provide a total order, unless all messages are required to start from the same node on the ring, which the Power5 does not. Rather, the requestor sends a request message around the ring and then absorbs the request when it sees it arrive back after traveling the entire ring. Each node observes the request on the ring and every processor in the node determines its snoop response. The first node to observe the request provides a single snoop response that is the aggregated snoop response of all of the processors on that node. A snoop response is *not* an actual data response, but rather a description of the action the chip or node would take. Without a totally ordered network, the chips/nodes cannot immediately act because they might not make consistent decisions about how to respond. The snoop response travels on the snoop response ring to the next node. This node similarly produces a single snoop response that aggregates the snoop response of the first node plus the snoop responses of all processors on the second node. When the aggregated snoop response of all nodes reaches the requestor chip, the requestor chip determines what every processor should do to respond to the request. The requestor chip broadcasts this decision along the ring to every node. This decision message is processed by every node/chip in the ring, and the node/chip that has been determined to be the one to provide a data response sends that data response on the data ring to the requestor.

This protocol is far more complicated than typical snooping protocols because of the lack of a totally ordered interconnection network. The protocol still has a total logical order of coherence requests, but without a totally ordered network, a node cannot immediately respond to a request because the request's position in the total order has not yet been determined by when it appears on the network. Despite the complexity, the Power5 design offers the advantages of having only point-to-point links and the simplicity of the ring topology (e.g., routing in a ring is so simple that switching can be faster than for other topologies). There have been other protocols that have exploited ring topologies and explored ordering issues for rings [2, 4, 7].

Extra Variants of Coherence States

The Power5 protocol is fundamentally a MESI protocol, but it has several "flavors" of some of these states. We list all of the states in Table 7.22. There are two new states that we wish to highlight. First, there is the SL variant of the Shared state. If an L2 cache holds a block in state SL, it may respond with data to a GetS from a processor on the same node, thus reducing this transaction's latency and reducing off-chip bandwidth demand; this ability to provide data distinguishes SL from S.

The other interesting new state is the T(agged) state. A block enters the T state when it is in Modified and receives a GetS request. Instead of downgrading to S, which it would do in a MESI protocol, or O, which it would do in a MOSI protocol, the cache changes the state to T. A block in state T is similar to the O state, in that it has a value that is more recent than the value in memory, and there may or may not be copies of the block in state S in other caches. Like the O state, the block may be read in the T state. Surprisingly, the T state is sometimes described as being a read-write state, which violates the SWMR invariant. Indeed, a store to state T may be performed immediately, and thus indeed violates the SWMR invariant in real (physical) time. However, the protocol still enforces the SWMR invariant in a logical time based on ring-order. Although the details of this ordering are beyond the scope of this primer, we think it helps to think of the T state as a variation on the E state. Recall that the E state allows

Table 7.22: Power5 L2 cache coherence states

State	Permissions	Description
I	None	Invalid
S	Read-only	Shared
SL	**Read-only**	**Shared local data source, but can respond with data to requests from processors in same node (sometimes referred to as F state, as in Intel QuickPath protocol (Section 8.8.4))**
S(S)	Read-only	Shared
Me (E)	Read-write	Exclusive
M (M)	Read-write	Modified
Mu	**Read-write**	**Modified unsolicited—received read-write data in response to read-only request**
T	Read-only	**Tagged—was M, received GetS. T is sometimes described as being a read-write state, which violates the SWMR invariant since there are also blocks in state S. A better way to think of T is that it is like E: it can immediately transition to M. However, unlike E, this transition is not silent: a store to a block in T state immediately transitions to M but (atomically) issues an invalidation message on the ring. Although other caches may race with this request, the T state has priority, and thus is guaranteed to be ordered first and thus does not need to wait for the invalidations to complete.**

a silent transition to M; thus a store to a block in state E may be immediately performed, so long as the state (atomically) transitions to state M. The T state is similar; a store in state T immediately transitions to state M. However, because there may also be copies in state S, a store in state T also causes the immediate issue of an invalidation message on the ring. Other cores may be attempting to upgrade from I or S to M, but the T state acts as the coherence ordering point and thus has priority and need not wait for an acknowledgment. It is not clear that this protocol is sufficient to support strong memory consistency models such as SC and TSO; however, as we discussed in Chapter 5, the Power memory model is one of the weakest memory consistency models. This Tagged state optimizes the common scenario of producer-consumer sharing, in which one thread writes a block and one or more other threads then read that block. The producer can re-obtain read-write access without having to wait as long each time.

7.8   DISCUSSION AND THE FUTURE OF SNOOPING

Snooping systems were prevalent in early multiprocessors because of their reputed simplicity and because their lack of scalability did not matter for the relatively small systems that dominated the market. Snooping also offers performance advantages for non-scalable systems because every snooping transaction can be completed with two messages, which we will contrast against the three-message transactions of directory protocols.

Despite its advantages, snooping is no longer commonly used. Even for small-scale systems, where snooping's lack of scalability is not a concern, snooping is no longer common. Snooping's requirement of a totally ordered broadcast network is just too costly, compared to the low-cost interconnection networks that suffice for directory protocols. Furthermore, for scalable systems, snooping is clearly a poor fit. Systems with very large numbers of cores are likely to be bottlenecked by both the interconnection network bandwidth needed to broadcast requests and the coherence controller bandwidth required to snoop every request. For such systems, a more scalable coherence protocol is required, and it is this need for scalability that originally motivated the directory protocols we present in the next chapter.

Snooping could be part of the solution though, even for scalable systems. As we will see in Section 9.1.6, one powerful technique for tackling scale is to divide and conquer. For example, a system consisting of multiple multicore chips could be kept coherent using a snooping coherence protocol within a chip and a scalable directory protocol across the chips.

7.9   REFERENCES

[1] N. Agarwal, L.-S. Peh, and N. K. Jha. In-network snoop ordering (INSO): Snoopy coherence on unordered interconnects. In *Proc. of the 14th International Symposium on High-Performance Computer Architecture*, pp. 67–78, February 2009. DOI: 10.1109/hpca.2009.4798238. 143

[2] L. A. Barroso and M. Dubois. Cache coherence on a slotted ring. In *Proc. of the 20th International Conference on Parallel Processing*, August 1991. 146

[3] A. Charlesworth. Starfire: Extending the SMP envelope. *IEEE Micro*, 18(1):39–49, January/February 1998. DOI: 10.1109/40.653032. 143, 144

[4] S. Frank, H. Burkhardt, III, and J. Rothnie. The KSR1: Bridging the gap between shared memory and MPPs. In *Proc. of the 38th Annual IEEE Computer Society Computer Conference (COMPCON)*, pp. 285–95, February 1993. DOI: 10.1109/cmpcon.1993.289682. 146

[5] M. Galles and E. Williams. Performance optimizations, implementation, and verification of the SGI Challenge multiprocessor. In *Proc. of the Hawaii International Conference on System Sciences*, 1994. DOI: 10.1109/hicss.1994.323177. 114

[6] M. M. K. Martin, D. J. Sorin, A. Ailamaki, A. R. Alameldeen, R. M. Dickson, C. J. Mauer, K. E. Moore, M. Plakal, M. D. Hill, and D. A. Wood. Timestamp snooping: An approach for extending SMPs. In *Proc. of the 9th International Conference on Architectural Support for Programming Languages and Operating Systems*, pp. 25–36, November 2000. DOI: 10.1145/378993.378998. 143

[7] M. R. Marty and M. D. Hill. Coherence ordering for ring-based chip multiprocessors. In *Proc. of the 39th Annual IEEE/ACM International Symposium on Microarchitecture*, December 2006. DOI: 10.1109/micro.2006.14. 146

[8] B. Sinharoy, R. N. Kalla, J. M. Tendler, R. J. Eickemeyer, and J. B. Joyner. POWER5 system microarchitecture. *IBM Journal of Research and Development*, 49(4/5), July/September 2005. DOI: 10.1147/rd.494.0505. 145

CHAPTER 8

Directory Coherence Protocols

In this chapter, we present directory coherence protocols. Directory protocols were originally developed to address the lack of scalability of snooping protocols. Traditional snooping systems broadcast all requests on a totally ordered interconnection network and all requests are snooped by all coherence controllers. By contrast, directory protocols use a level of indirection to avoid both the ordered broadcast network and having each cache controller process every request.

We first introduce directory protocols at a high level (Section 8.1). We then present a system with a complete but unsophisticated three-state (MSI) directory protocol (Section 8.2). This system and protocol serve as a baseline upon which we later add system features and protocol optimizations. We then explain how to add the Exclusive state (Section 8.3) and the Owned state (Section 8.4) to the baseline MSI protocol. Next we discuss how to represent the directory state (Section 8.5) and how to design and implement the directory itself (Section 8.6). We then describe techniques for improving performance and reducing the implementation costs (Section 8.7). We then discuss commercial systems with directory protocols (Section 8.8) before concluding the chapter with a discussion of directory protocols and their future (Section 8.9).

Those readers who are content to learn just the basics of directory coherence protocols can skim or skip Section 8.3 through Section 8.7, although some of the material in these sections will help the reader to better understand the case studies in Section 8.8.

8.1 INTRODUCTION TO DIRECTORY PROTOCOLS

The key innovation of directory protocols is to establish a *directory* that maintains a global view of the coherence state of each block. The directory tracks which caches hold each block and in what states. A cache controller that wants to issue a coherence request (e.g., a GetS) sends it directly to the directory (i.e., a unicast message), and the directory looks up the state of the block to determine what actions to take next. For example, the directory state might indicate that the requested block is owned by core C2's cache and thus the request should be forwarded to C2 (e.g., using a new Fwd-GetS request) to obtain a copy of the block. When C2's cache controller receives this forwarded request, it unicasts a response to the requesting cache controller.

It is instructive to compare the basic operation of directory protocols and snooping protocols. In a directory protocol, the directory maintains the state of each block, and cache controllers send all requests to the directory. The directory either responds to the request or forwards the request to one or more other coherence controllers that then respond. Coherence transactions typically involve either two steps (a unicast request, followed by a unicast response) or three steps (a unicast request, $K \geq 1$ forwarded requests, and K responses, where K is the number of sharers). Some protocols even have a fourth step, either because responses indirect through the directory or because the requestor notifies the directory on transaction completion. In contrast, snooping protocols distribute a block's state across potentially all of the coherence controllers. Because there is no central summary of this distributed state, coherence requests must be broadcast to all coherence controllers. Snooping coherence transactions thus always involve two steps (a broadcast request, followed by a unicast response).

Like snooping protocols, a directory protocol needs to define when and how coherence transactions become ordered with respect to other transactions. In most directory protocols, a coherence transaction is ordered at the directory. Multiple coherence controllers may send coherence requests to the directory at the same time, and the transaction order is determined by the order in which the requests are serialized at the directory. If two requests race to the directory, the interconnection network effectively chooses which request the directory will process first. The fate of the request that arrives second is a function of the directory protocol and what types of requests are racing. The second request might get (a) processed immediately after the first request, (b) held at the directory while awaiting the first request to complete, or (c) negatively acknowledged (NACKed). In the latter case, the directory sends a negative acknowledgment message (NACK) to the requestor, and the requestor must re-issue its request. In this chapter, we do not consider protocols that use NACKs, but we do discuss the possible use of NACKs and how they can cause livelock problems in Section 9.3.2.

Using the directory as the ordering point represents another key difference between directory protocols and snooping protocols. Traditional snooping protocols create a total order by serializing all transactions on the ordered broadcast network. Snooping's total order not only ensures that each block's requests are processed in per-block order but also facilitates implementing a memory consistency model. Recall that traditional snooping protocols use totally ordered broadcast to serialize all requests; thus, when a requestor observes its own coherence request this serves as notification that its coherence epoch may begin. In particular, when a snooping controller sees its own GetM request, it can infer that other caches will invalidate their S blocks. We demonstrated in Table 7.4 that this serialization notification is sufficient to support the strong SC and TSO memory consistency models.

In contrast, a directory protocol orders transactions at the directory to ensure that conflicting requests are processed by all nodes in per-block order. However, the lack of a total order means that a requestor in a directory protocol needs another strategy to determine when its request has been serialized and thus when its coherence epoch may safely begin. Because (most) directory protocols do not use totally ordered broadcast, there is no global notion of serialization. Rather, a request must be individually serialized with respect to all the caches that (may) have a copy of the block. Explicit messages are needed to notify the requestor that its request has been serialized by each relevant cache. In particular, on a GetM request, each cache controller with a shared (S) copy must send an explicit acknowledgment (Ack) message once it has serialized the invalidation message.

This comparison between directory and snooping protocols highlights the fundamental trade-off between them. A directory protocol achieves greater scalability (i.e., because it requires less bandwidth) at the cost of a level of indirection (i.e., having three steps, instead of two steps, for some transactions). This additional level of indirection increases the latency of some coherence transactions.

8.2   BASELINE DIRECTORY SYSTEM

In this section, we present a baseline system with a straightforward, modestly optimized directory protocol. This system provides insight into the key features of directory protocols while revealing inefficiencies that motivate the features and optimizations presented in subsequent sections of this chapter.

8.2.1   DIRECTORY SYSTEM MODEL

We illustrate our directory system model in Figure 8.1. Unlike for snooping protocols, the topology of the interconnection network is intentionally vague. It could be a mesh, torus, or any other topology that the architect wishes to use. One restriction on the interconnection network that we assume in this chapter is that it enforces point-to-point ordering. That is, if controller A sends two messages to controller B, then the messages arrive at controller B in the same order in which they were sent.[1] Having point-to-point ordering reduces the complexity of the protocol, and we defer a discussion of networks without ordering until Section 8.7.3.

The only differences between this directory system model and the baseline system model in Figure 2.1 is that we have added a directory and we have renamed the memory controller to be the directory controller. There are many ways of sizing and organizing the directory, and for now we assume the simplest model: for each block in memory, there is a corresponding directory entry. In Section 8.6, we examine and compare more practical directory organization options. We also assume a monolithic LLC with a single directory controller; in Section 8.7.1, we explain how to distribute this functionality across multiple banks of an LLC and multiple directory controllers.

8.2.2   HIGH-LEVEL PROTOCOL SPECIFICATION

The baseline directory protocol has only three stable states: MSI. A block is owned by the directory controller unless the block is in a cache in state M. The directory state for each block includes the stable coherence state, the identity of the owner (if the block is in state M), and the

---

[1]Strictly speaking, we require point-to-point order for only certain types of messages, but this is a detail that we defer until Section 8.7.3.

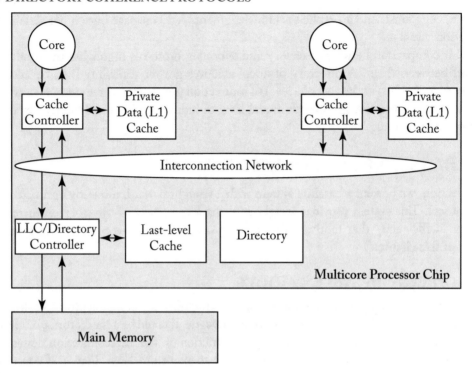

Figure 8.1: Directory system model.

identities of the sharers encoded as a one-hot bit vector (if the block is in state S). We illustrate a directory entry in Figure 8.2. In Section 8.5, we will discuss other encodings of directory entries.

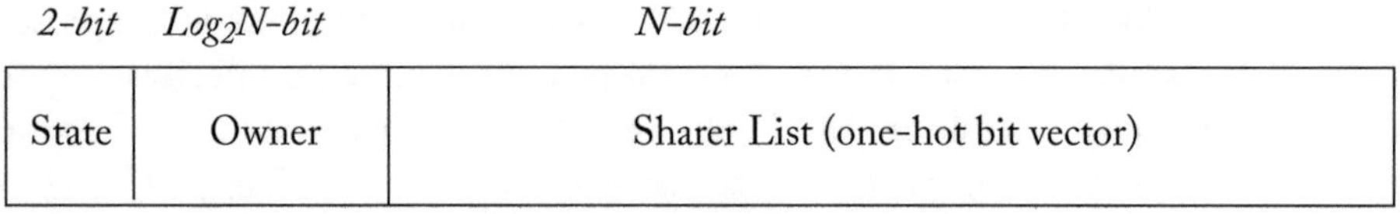

2-bit	Log_2N-bit	N-bit
State	Owner	Sharer List (one-hot bit vector)

Figure 8.2: Directory entry for a block in a system with N nodes.

Before presenting the detailed specification, we first illustrate a higher level abstraction of the protocol in order to understand its fundamental behaviors. In Figure 8.3, we show the transactions in which a cache controller issues coherence requests to change permissions from I to S, I or S to M, M to I, and S to I. As with the snooping protocols in the last chapter, we specify the directory state of a block using a cache-centric notation (e.g., a directory state of M denotes that there exists a cache with the block in state M). Note that a cache controller may not silently evict a Shared block; that is, there is an explicit PutS request. We defer a discussion of protocols with silent evictions of shared blocks, as well as a comparison of silent vs. explicit PutS requests, until Section 8.7.4.

Most of the transactions are fairly straightforward, but two transactions merit further discussion here. The first is the transaction that occurs when a cache is trying to upgrade permissions from I or S to M and the directory state is S. The cache controller sends a GetM to the directory, and the directory takes two actions. First, it responds to the requestor with a message that includes the data and the "AckCount;" the AckCount is the number of current sharers of the block. The directory sends the AckCount to the requestor to inform the requestor of how many sharers must acknowledge having invalidated their block in response to the GetM. Second, the directory sends an Invalidation (Inv) message to all of the current sharers. Each sharer, upon receiving the Invalidation, sends an Invalidation-Ack (Inv-Ack) to the requestor. Once the requestor receives the message from the directory and *all* of the Inv-Ack messages, it completes the transaction. The requestor, having received all of the Inv-Ack messages, knows that there are no longer any readers of the block and thus it may write to the block without violating coherence.

The second transaction that merits further discussion occurs when a cache is trying to evict a block in state M. In this protocol, we have the cache controller send a PutM message that includes the data to the directory. The directory responds with a Put-Ack. If the PutM did not carry the data with it, then the protocol would require a third message—a data message from the cache controller to the directory with the evicted block that had been in state M—to be sent in a PutM transaction. The PutM transaction in this directory protocol differs from what occurred in the snooping protocol, in which a PutM did not carry data.

8.2.3   AVOIDING DEADLOCK

In this protocol, the reception of a message can cause a coherence controller to send another message. In general, if event A (e.g., message reception) can cause event B (e.g., message sending) and both these events require resource allocation (e.g., network links and buffers), then we must be careful to avoid deadlock that could occur if circular resource dependences arise. For example, a GetS request can cause the directory controller to issue a Fwd-GetS message; if these messages use the same resources (e.g., network links and buffers), then the system can potentially deadlock. In Figure 8.4, we illustrate a deadlock in which two coherence controllers C1 and C2 are responding to each other's requests, but the incoming queues are already full of other coherence requests. If the queues are FIFO, then the responses cannot pass the requests. Because the queues are full, each controller stalls trying to send a response. Because the queues are FIFO, the controller cannot switch to work on a subsequent request (or get to the response). Thus, the system deadlocks.

A well-known solution for avoiding deadlock in coherence protocols is to use separate networks for each class of message. The networks can be physically separate or logically separate (called *virtual networks*), but the key is avoiding dependences between classes of messages.

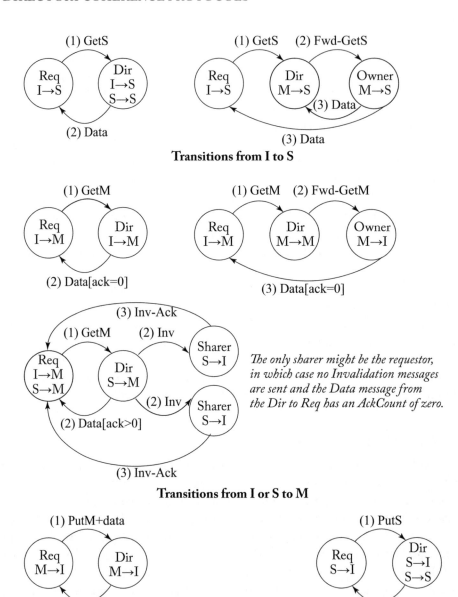

Figure 8.3: High-level description of MSI directory protocol. In each transition, the cache controller that requests the transaction is denoted "Req."

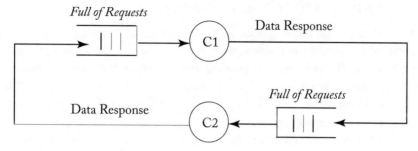

Figure 8.4: Deadlock example.

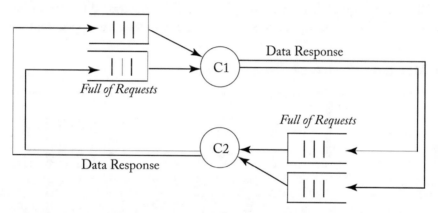

Figure 8.5: Avoiding deadlock with separate networks.

Figure 8.5 illustrates a system in which request and response messages travel on separate physical networks. Because a response cannot be blocked by another request, it will eventually be consumed by its destination node, breaking the cyclic dependence.

The directory protocol in this section uses three networks to avoid deadlock. Because a request can cause a forwarded request and a forwarded request can cause a response, there are three message classes that each require their own network. Request messages are GetS, GetM, and PutM. Forwarded request messages are Fwd-GetS, Fwd-GetM, Inv(alidation), and Put-Ack. Response messages are Data and Inv-Ack. The protocols in this chapter require that the Forwarded Request network provides point-to-point ordering; other networks have no ordering constraints nor are there any ordering constraints between messages traveling on different networks.

We defer a more thorough discussion of deadlock avoidance, including more explanation of virtual networks and the exact requirements for avoiding deadlock, until Section 9.3.

8.2.4   DETAILED PROTOCOL SPECIFICATION

We present the detailed protocol specification, including all transient states, in Tables 8.1 and 8.2. Compared to the high-level description in Section 8.2.2, the most significant difference is the transient states. The coherence controllers must manage the states of blocks that are in the midst of coherence transactions, including situations in which a cache controller receives a forwarded request from another controller in between sending its coherence request to the directory and receiving all of its necessary response messages, including Data and possible Inv-Acks. The cache controllers can maintain this state in the miss status handling registers (MSHRs) that cores use to keep track of outstanding coherence requests. Notationally, we represent these transient states in the form XY^{AD}, where the superscript A denotes waiting for acknowledgments and the superscript D denotes waiting for data. (This notation differs from the snooping protocols, in which the superscript A denoted waiting for a request to appear on the bus.)

Table 8.1: MSI directory protocol—cache controller

	Load	Store	Replacement	Fwd-GetS	Fwd-GetM	Inv	Put-Ack	Data from Dir (ack=0)	Data from Dir (ack>0)	Data from Owner	Inv-Ack	Last-Inv-Ack
I	Send GetS to Dir/IS^D	Send GetM to Dir/IM^{AD}										
IS^D	Stall	Stall	Stall			Stall		-/S			-/S	
IM^{AD}	Stall	Stall	Stall	Stall	Stall			-/M	-/IM^A	-/M	ack--	
IM^A	Stall	Stall	Stall	Stall	Stall						ack--	-/M
S	Hit	Send GetM to Dir/SM^{AD}	Send PutS to Dir/SI^A			Send Inv-Ack to Req/I						
SM^{AD}	Hit	Stall	Stall	Stall	Stall	Send Inv-Ack to Req/IM^{AD}		-/M	-/SM^A		ack--	
SM^A	Hit	Stall	Stall	Stall	Stall						ack--	-/M
M	Hit	Hit	Send PutM +data to Dir/MI^A	Send data to Req and Dir/S	Send data to Req/I							
MI^A	Stall	Stall	Stall	Send data to Req and Dir/SI^A	Send data to Req/II^A		-/I					
SI^A	Stall	Stall	Stall			Send Inv-Ack to Req/II^A	-/I					
II^A	Stall	Stall	Stall				-/I					

Table 8.2: MSI directory protocol—directory controller

	GetS	GetM	PutS-NotLast	PutS-Last	Put M+ data from Owner	PutM+data from Non-Owner	Data
I	Send data to Req, add Req to Sharers/S	Send data to Req, set Owner to Req/M	Send Put-Ack to Req	Send Put-Ack to Req		Send Put-Ack to Req	
S	Send data to Req, add Req to Sharers	Send data to Req, send Inv to Sharers, clear Sharers, set Owner to Req/M	Remove Req from Sharers, sent Put-Ack to Req	Remove Req from Sharers, send Put-Ack to Req/I		Remove Req from sharers, send Put-Ack to Req	
M	Send Fwd-GetS to Owner, add Req and Owner to Sharers, clear Owner/S^D	Send Fwd-GetM to Owner, set Owner to Req	Send Put-Ack to Req	Send Put-Ack to Req	Copy data to memory, clear Owner, send Put-Ack to Req/I	Send Put-Ack to Req	
S^D	Stall	Stall	Remove Req from Sharers, send Put-Ack to Req	Remove Req from Sharers, send Put-Ack to Req		Remove Req from Sharers, send Put-Ack to Req	Copy data to memory/S

Because these tables can be somewhat daunting at first glance, the next section walks through some example scenarios.

8.2.5 PROTOCOL OPERATION

The protocol enables caches to acquire blocks in states S and M and to replace blocks to the directory in either of these states.

I to S (common case #1)

The cache controller sends a GetS request to the directory and changes the block state from I to IS^D. The directory receives this request and, if the directory is the owner (i.e., no cache currently has the block in M), the directory responds with a Data message, changes the block's state to S (if it is not S already), and adds the requestor to the sharer list. When the Data arrives at the requestor, the cache controller changes the block's state to S, completing the transaction.

I to S (common case #2)

The cache controller sends a GetS request to the directory and changes the block state from I to IS^D. If the directory is *not* the owner (i.e., there is a cache that currently has the block in M), the directory forwards the request to the owner and changes the block's state to the transient state S^D. The owner responds to this Fwd-GetS message by sending Data to the requestor and changing the block's state to S. The now-previous owner must also send Data to the directory since it is relinquishing ownership to the directory, which must have an up-to-date copy of the block. When the Data arrives at the requestor, the cache controller changes the block state to S and considers the transaction complete. When the Data arrives at the directory, the directory copies it to memory, changes the block state to S, and considers the transaction complete.

I to S (race cases)

The above two I-to-S scenarios represent the common cases, in which there is only one transaction for the block in progress. Most of the protocol's complexity derives from having to deal with the less-common cases of multiple in-progress transactions for a block. For example, a reader may find it surprising that a cache controller can receive an Invalidation for a block in state IS^D. Consider core C1 that issues a GetS and goes to IS^D and another core C2 that issues a GetM for the same block that arrives at the directory after C1's GetS. The directory first sends C1 Data in response to its GetS and then an Invalidation in response to C2's GetM. Because the Data and Invalidation travel on separate networks, they can arrive out of order, and thus C1 can receive the Invalidation before the Data.

I or S to M

The cache controller sends a GetM request to the directory and changes the block's state from I to IM^{AD}. In this state, the cache waits for Data and (possibly) Inv-Acks that indicate that other caches have invalidated their copies of the block in state S. The cache controller knows how many Inv-Acks to expect, since the Data message contains the AckCount, which may be zero. Figure 8.3 illustrates the three common-case scenarios of the directory responding to the GetM request. If the directory is in state I, it simply sends Data with an AckCount of zero and goes to state M. If in state M, the directory controller forwards the request to the owner and updates the block's owner; the now-previous owner responds to the Fwd-GetM request by sending Data with an AckCount of zero. The last common case occurs when the directory is in state S. The directory responds with Data and an AckCount equal to the number of sharers, plus it sends Invalidations to each core in the sharer list. Cache controllers that receive Invalidation messages invalidate their shared copies and send Inv-Acks to the requestor. When the requestor receives the last Inv-Ack, it transitions to state M. Note the special Last-Inv-Ack event in Table 8.1, which simplifies the protocol specification.

These common cases neglect some possible races that highlight the concurrency of directory protocols. For example, core C1 has the cache block in state IM^A and receives a Fwd-GetS from C2's cache controller. This situation is possible because the directory has already sent Data to C1, sent Invalidation messages to the sharers, and changed its state to M. When C2's GetS arrives at the directory, the directory simply forwards it to the owner, C1. This Fwd-GetS may arrive at C1 before all of the Inv-Acks arrive at C1. In this situation, our protocol simply stalls and the cache controller waits for the Inv-Acks. Because Inv-Acks travel on a separate network, they are guaranteed not to block behind the unprocessed Fwd-GetS.

M to I

To evict a block in state M, the cache controller sends a PutM request that includes the data and changes the block state to MI^A. When the directory receives this PutM, it updates the LLC/memory, responds with a Put-Ack, and transitions to state I. Until the requestor receives the Put-Ack, the block's state remains effectively M and the cache controller must respond to forwarded coherence requests for the block. In the case where the cache controller receives a forwarded coherence request (Fwd-GetS or Fwd-GetM) between sending the PutM and receiving the Put-Ack, the cache controller responds to the Fwd-GetS or Fwd-GetM and changes its block state to SI^A or II^A, respectively. These transient states are effectively S and I, respectively, but denote that the cache controller must wait for a Put-Ack to complete the transition to I.

S to I

Unlike the snooping protocols in the previous chapter, our directory protocols do not silently evict blocks in state S. Instead, to replace a block in state S, the cache controller sends a PutS request and changes the block state to SI^A. The directory receives this PutS and responds with a Put-Ack. Until the requestor receives the Put-Ack, the block's state is effectively S. If the cache controller receives an Invalidation request after sending the PutS and before receiving the Put-Ack, it changes the block's state to II^A. This transient state is effectively I, but it denotes that the cache controller must wait for a Put-Ack to complete the transaction from S to I.

8.2.6   PROTOCOL SIMPLIFICATIONS

This protocol is relatively straightforward and sacrifices some performance to achieve this simplicity. We now discuss two simplifications:

- The most significant simplification, other than having only three stable states, is that the protocol stalls in certain situations. For example, a cache controller stalls when it receives a forwarded request while in a transient state. A higher performance option, discussed in Section 8.7.2, would be to process the messages and add more transient states.

- A second simplification is that the directory sends Data (and the AckCount) in response to a cache that is changing a block's state from S to M. The cache already has valid data and thus it would be sufficient for the directory to simply send a data-less AckCount. We defer adding this new type of message until we present the MOSI protocol in Section 8.4.

8.3   ADDING THE EXCLUSIVE STATE

As we previously discussed in the context of snooping protocols, adding the Exclusive (E) state is an important optimization because it enables a core to read and then write a block with only a single coherence transaction, instead of the two required by an MSI protocol. At the highest level, this optimization is independent of whether the cache coherence uses snooping or directories. If a core issues a GetS and the block is not currently shared by other cores, then the requestor may obtain the block in state E. The core may then silently upgrade the block's state from E to M without issuing another coherence request.

In this section, we add the E state to our baseline MSI directory protocol. As with the MESI snooping protocol in the previous chapter, the operation of the protocol depends on whether the E state is considered an ownership state or not. And, as with the MESI snooping protocol, the primary operational difference involves determining which coherence controller should respond to a request for a block that the directory gave to a cache in state E. The block may have been silently upgraded from E to M since the directory gave the block to the cache in state E.

In protocols in which an E block is owned, the solution is simple. The cache with the block in E (or M) is the owner and thus must respond to requests. A coherence request sent to the directory will be forwarded to the cache with the block in state E. Because the E state is an ownership state, the eviction of an E block cannot be performed silently; the cache must issue a PutE request to the directory. Without an explicit PutE, the directory would not know that the directory was now the owner and should respond to incoming coherence requests. Because we assume in this primer that blocks in E are owned, this simple solution is what we implement in the MESI protocol in this section.

In protocols in which an E block is not owned, an E block can be silently evicted, but the protocol complexity increases. Consider the case where core C1 obtains a block in state E and then the directory receives a GetS or GetM from core C2. The directory knows that C1 is either (i) still in state E, (ii) in state M (if C1 did a store with a silent upgrade from E to M), or (iii) in state I (if the protocol allows C1 to perform a silent PutE). If C1 is in M, the directory must forward the request to C1 so that C1 can supply the latest version of the data. If C1 is in E, C1 or the directory may respond since they both have the same data. If C1 is in I, the directory must respond. One solution, which we describe in more detail in our case study on the SGI Origin [10] in Section 8.8.1, is to have both C1 and the directory respond. Another solution is to have the directory forward the request to C1. If C1 is in I, C1 notifies the directory to respond to C2; else, C1 responds to C2 and notifies the directory that it does not need to respond to C2.

8.3.1   HIGH-LEVEL PROTOCOL SPECIFICATION

We specify a high-level view of the transactions in Figure 8.6, with differences from the MSI protocol highlighted. There are only two significant differences. First, there is a transition from I

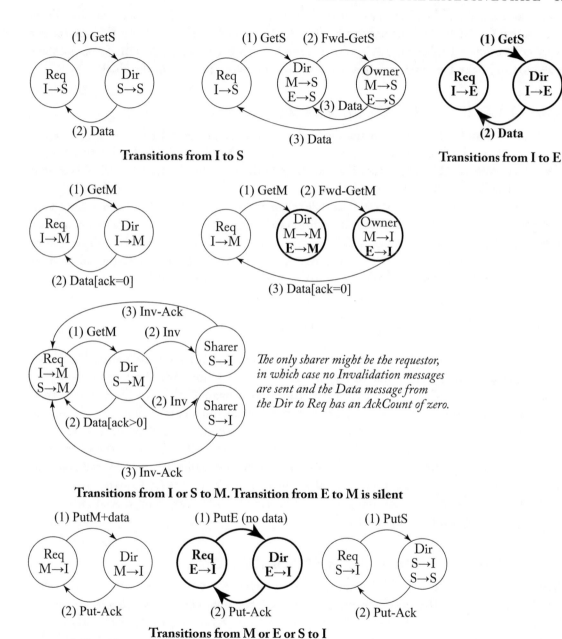

Figure 8.6: High-level description of MESI directory protocol. In each transition, the cache controller that requests the transaction is denoted "Req."

to E that can occur if the directory receives a GetS for a block in state I. Second, there is a PutE transaction for evicting blocks in state E. Because E is an ownership state, an E block cannot be silently evicted. Unlike a block in state M, the E block is clean, and thus the PutE does not need to carry data; the directory already has the most up-to-date copy of the block.

8.3.2   DETAILED PROTOCOL SPECIFICATION

In Tables 8.3 and 8.4, we present the detailed specification of the MESI protocol, including transient states. Differences with respect to the MSI protocol are highlighted with boldface font. The protocol adds to the set of cache states both the stable E state as well as transient states to handle transactions for blocks initially in state E.

This protocol is somewhat more complex than the MSI protocol, with much of the added complexity at the directory controller. In addition to having more states, the directory controller must distinguish between more possible events. For example, when a PutS arrives, the directory must distinguish whether this is the "last" PutS; that is, did this PutS arrive from the only current sharer? If this PutS is the last PutS, then the directory's state changes to I.

8.4   ADDING THE OWNED STATE

For the same reason we added the Owned state to the baseline MSI snooping protocol in Chapter 7, an architect may want to add the Owned state to the baseline MSI directory protocol presented in Section 8.2. Recall from Chapter 2 that if a cache has a block in the Owned state, then the block is valid, read-only, dirty (i.e., it must eventually update memory), and owned (i.e., the cache must respond to coherence requests for the block). Adding the Owned state changes the protocol, compared to MSI, in three important ways: (1) a cache with a block in M that observes a Fwd-GetS changes its state to O and does not need to (immediately) copy the data back to the LLC/memory, (2) more coherence requests are satisfied by caches (in O state) than by the LLC/memory, and (3) there are more 3-hop transactions (which would have been satisfied by the LLC/memory in an MSI protocol).

8.4.1   HIGH-LEVEL PROTOCOL SPECIFICATION

We specify a high-level view of the transactions in Figure 8.7, with differences from the MSI protocol highlighted. The most interesting difference is the transaction in which a requestor of a block in state I or S sends a GetM to the directory when the block is in the O state in the owner cache and in the S state in one or more sharer caches. In this case, the directory forwards the GetM to the owner, and appends the AckCount. The directory also sends Invalidations to each of the sharers. The owner receives the Fwd-GetM and responds to the requestor with Data and the AckCount. The requestor uses this received AckCount to determine when it has received the last Inv-Ack. There is a similar transaction if the requestor of the GetM was the owner (in

Table 8.3: MESI directory protocol—cache controller

	Load	Store	Replacement	Fwd-GetS	Fwd-GetM	Inv	Put-Ack	Exclusive Data from Dir	Data from Dir (ack=0)	Data from Dir (ack>0)	Data from Owner	Inv-Ack	Last-Inv-Ack
I	Sent GetS to Dir/IS^D	Send GetM to Dir/IM^{AD}											
IS^D	Stall	Stall	Stall			Stall		-/E	-/S		-/S		
IM^{AD}	Stall	Stall	Stall	Stall	Stall				-/M	-/IM^A	-/M	ack--	
IM^A	Stall	Stall	Stall	Stall	Stall							ack--	-/M
S	Hit	Send GetM to Dir/SM^{AD}	Send PutS to Dir/SI^A			Send Inv-Ack to Req/I							
SM^{AD}	Hit	Stall	Stall	Stall	Stall	Send Inv-Ack to Req/ IM^{AD}			-/M	-/SM^A		ack--	
SM^A	Hit	Stall	Stall	Stall	Stall							ack--	-/M
M	Hit	Hit	Send PutM +data to Dir/MI^A	Send data to Req and Dir/S	Send data to Req/I								
E	**Hit**	**Hit/M**	**Send PutE (no data) to Dir/ EI^A**	**Send data to Req and Dir/S**	**Send data to Req/I**								
MI^A	Stall	Stall	Stall	Send data to Req and Dir/SI^A	Send data to Req/II^A		-/I						
EI^A	**Stall**	**Stall**	**Stall**	**Send data to Req and Dir/SI^A**	**Send data to Req/ II^A**		-/I						
SI^A	Stall	Stall	Stall			Send Inv-Ack to Req/II^A	-/I						
II^A	Stall	Stall	Stall				-/I						

Table 8.4: MESI directory protocol—directory controller

	GetS	GetM	PutS-NotLast	PutS-Last	PutM+data from Owner	PutM from Non-Owner	PutE (no-data) from Owner	PutE from Non-Owner	Data
I	**Send Exclusive data to Req, set Owner to Req/E**	Send data to Req, set Owner to Req/M	Sent Put-Ack to Req	Send Put-Ack to Req		Send Put-Ack to Req		**Send Put-Ack to Req**	
S	Send data to Req, add Req to Sharers	Send data to Req, send Inv to Sharers, clear Sharers, set Owner to Req/M	Remove Req from Sharers, send Put-Ack to Req	Remove Req from Sharers, send Put-Ack to Req/I		Remove Req from Sharers, send Put-Ack to Req		**Remove Req from Sharers, send Put-Ack to Req**	
E	**Forward GetS to Owner, make Owner sharer, add Req to Sharers, clear Owner /S^D**	**Forward GetM to Owner, set Owner to Req/M**	**Send Put-Ack to Req**	**Send Put-Ack to Req**	**Copy data to mem, send Put-Ack to Req, clear Owner/I**	**Send Put-Ack to Req**	**Send Put-Ack to Req, clear Owner/I**	**Send Put-Ack to Req**	
M	Forward GetS to Owner, make Owner sharer, add Req to Sharers, clear Owner /S^D	Forward GetM to Owner, set Owner to Req	Sent Put-Ack to Req	Sent Put-Ack to Req	Copy data to mem, send Put-Ack to Req, clear Owner/I	Send Put-Ack to Req		**Send Put-Ack to Req**	
S^D	Stall	Stall	Remove Req from Sharers, send Put-Ack to Req	Remove Req from Sharers, send Put-Ack to Req		Remove Req from Sharers, send Put-Ack to Req		**Remove Req from Sharers, send Put-Ack to Req**	Copy data to LLC/mem/S

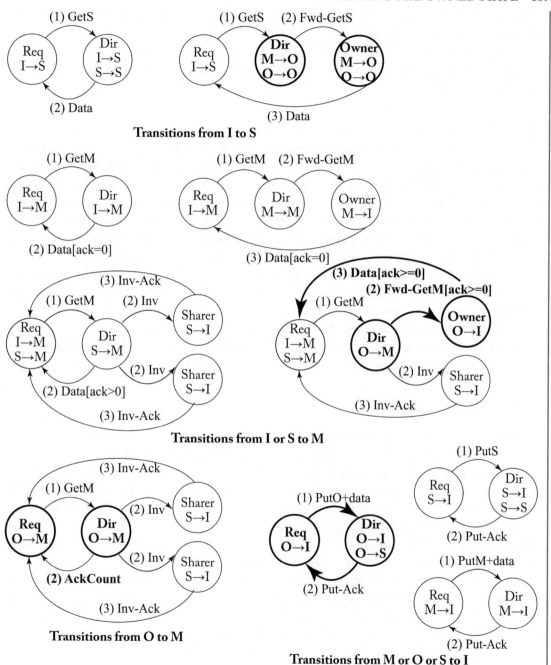

Figure 8.7: High-level description of MOSI directory protocol. In each transition, the cache controller that requests the transaction is denoted "Req."

state O). The difference here is that the directory sends the AckCount directly to the requestor because the requestor is the owner.

This protocol has a PutO transaction that is nearly identical to the PutM transaction. It contains data for the same reason that the PutM transaction contains data, i.e., because both M and O are dirty states.

8.4.2   DETAILED PROTOCOL SPECIFICATION

Tables 8.5 and 8.6 present the detailed specification of the MOSI protocol, including transient states. Differences with respect to the MSI protocol are highlighted with boldface font. The protocol adds to the set of cache states both the stable O state as well as transient OM^{AC}, OM^A, and OI^A states to handle transactions for blocks initially in state O. The state OM^{AC} indicates that the cache is waiting for both Inv-Acks (A) from caches and an AckCount (C) from the directory, but not data. Because this block started in state O, it already had valid data.

An interesting situation arises when core C1's cache controller has a block in OM^{AC} or SM^{AD} and receives a Fwd-GetM or Invalidation from core C2 for that block. C2's GetM must have been ordered at the directory before C1's GetM, for this situation to arise. Thus, the directory state changes to M (owned by C2) before observing C1's GetM. When C2's Fwd-GetM or Invalidation arrives at C1, C1 must be aware that C2's GetM was ordered first. Thus, C1's cache state changes from either OM^{AC} or SM^{AD} to IM^{AD}. The forwarded GetM or Invalidation from C2 invalidated C1's cache block and now C1 must wait for both Data and Inv-Acks.

8.5   REPRESENTING DIRECTORY STATE

In the previous sections, we have assumed a *complete directory*; that is, the directory maintains the complete state for each block, including the full set of caches that (may) have shared copies. Yet this assumption contradicts the primary motivation for directory protocols: scalability. In a system with a large number of caches (i.e., a large number of potential sharers of a block), maintaining the complete set of sharers for each block requires a significant amount of storage, even when using a compact bit-vector representation. For a system with a modest number of caches, it may be reasonable to maintain this complete set, but the architects of larger-scale systems may wish for more scalable solutions to maintaining directory state.

There are many ways to reduce how much state the directory maintains for each block. Here we discuss two important techniques: coarse directories and limited pointers. We discuss these techniques independently, but observe that they can be combined. We contrast each solution with the baseline design, illustrated in the top entry of Figure 8.8.

8.5.1   COARSE DIRECTORY

Having the complete set of sharers enables the directory to send Invalidation messages to exactly those cache controllers that have the block in state S. One way to reduce the directory state is

Table 8.5: MOSI directory protocol—cache controller

	Load	Store	Replacement	Fwd-GetS	Fwd-GetM	Inv	Put-Ack	Data from Dir (ack=0)	Data from Dir (ack>0)	Data from Owner	AckCount from Dir	Inv-Ack	Last-Inv-Ack
I	Sent GetS to Dir/IS^D	Send GetM to Dir/IM^{AD}											
IS^D	Stall	Stall	Stall			Stall	-/S		-/S				
IM^{AD}	Stall	Stall	Stall	Stall	Stall			-/M	-/IM^A	-/M		ack--	
IM^A	Stall	Stall	Stall	Stall	Stall							ack--	-/M
S	Hit	Send GetM to Dir/SM^{AD}	Send PutS to Dir/SI^A			Send Inv-Ack to Req/I							
SM^{AD}	Hit	Stall	Stall	Stall	Stall	Send Inv-Ack to Req/IM^{AD}		-/M	-/SM^A			ack--	
SM^A	Hit	Stall	Stall	Stall	Stall							ack--	-/M
M	Hit	Hit	Send PutM +data to Dir/MI^A	**Send data to Req/O**	Send data to Req/I								
MI^A	Stall	Stall	Stall	**Send data to Req/OI^A**	Send data to Req/II^A		-/I						
O	**Hit**	**Send GetM to Dir/OM^{AC}**	**Send PutO +data to Dir/OI^A**	**Send data to Req**	**Send data to Req/I**								
OM^{AC}	**Hit**	**Stall**	**Stall**	**Send data to Req**	**Send data to Req/IM^{AD}**						-/OM^A	ack--	
OM^A	Hit	**Stall**	**Stall**	**Stall**	**Stall**							ack--	-/M
OI^A	**Stall**	**Stall**	**Stall**	**Send data to Req**	**Send data to Req/II^A**		-/I						
SI^A	Stall	Stall	Stall			Send Inv-Ack to Req/II^A	-/I						
II^A	Stall	Stall	Stall				-/I						

Table 8.6: **MOSI** directory protocol—directory controller

	GetS	GetM from Owner	GetM from NonOwner	PutS-NotLast	PutS-Last	PutM+data from Owner	PutM+data from Non-Owner	PutO-data from Owner	PutO-data from Non-Owner
I	Send data to Req, add Req to Sharer/S		Send data to Req, set Owner to Req/M	Send Put-Ack to Req	Send Put-Ack to Req		Send Put-Ack to Req		**Send Put-Ack to Req**
S	Send data to Req, add Req to Sharers		Send data to Req, send Inv to Sharers, set Owner to Req, clear Sharers/M	Remove Req from Sharers, send Put-Ack to Req	Remove Req from Sharers, send Put-Ack to Req/I		Remove Req from Sharers, send Put-Ack to Req		**Remove Req from Sharers, send Put-Ack to Req**
O	**Forward GetS to Owner, add Req to Sharers**	**Send Ack-Count to Req, send Inv to Sharers, clear Sharers/M**	**Forward GetM to Owner, send Inv to Sharers, set Owner to Req, clear Sharers, send Ack-Count to Req/M**	**Remove Req from Sharers, send Put-Ack to Req**	**Remove Req from Sharers, send Put-Ack to Req**	**Remove Req from Sharers, copy data to mem, send Put-Ack to Req, clear Owner/S**	**Remove Req from Sharers, send Put-Ack to Req**	**Copy data to mem, send Put-Ack to Req, clear Owner/S**	**Remove Req from Sharers, send Put-Ack to Req**
M	**Forward GetS to Owner, add Req to Sharers/O**		Forward GetM to Owner, set Owner to Req	Sent Put-Ack to Req	Sent Put-Ack to Req	Copy data to mem, send Put-Ack to Req, clear Owner/I	Sent Put-Ack to Req		**Sent Put-Ack to Req**

to conservatively maintain a coarse list of sharers that is a superset of the actual set of sharers. That is, a given entry in the sharer list corresponds to a set of K caches, as illustrated in the middle entry of Figure 8.8. If one or more of the caches in that set (may) have the block in state S, then that bit in the sharer list is set. A GetM will cause the directory controller to send an Invalidation to all K caches in that set. Thus, coarse directories reduce the directory state at the expense of extra inter-connection network bandwidth for unnecessary Invalidation messages, plus the cache controller bandwidth to process these extra Invalidation messages.

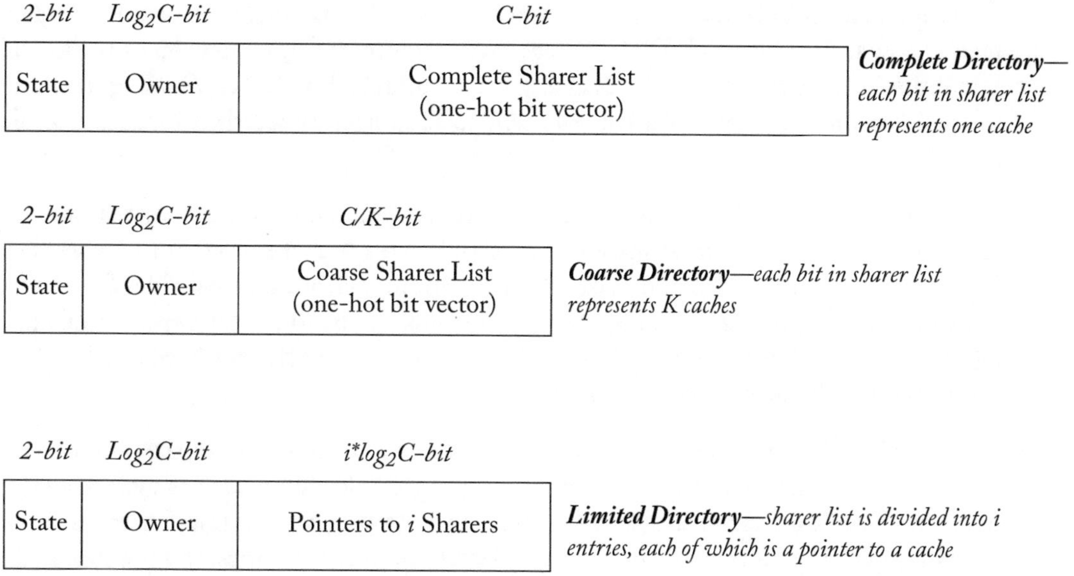

2-bit	*Log_2C-bit*	*C-bit*	
State	Owner	Complete Sharer List (one-hot bit vector)	**Complete Directory—** *each bit in sharer list represents one cache*

2-bit	*Log_2C-bit*	*C/K-bit*	
State	Owner	Coarse Sharer List (one-hot bit vector)	**Coarse Directory**—*each bit in sharer list represents K caches*

2-bit	*Log_2C-bit*	*$i*log_2C$-bit*	
State	Owner	Pointers to *i* Sharers	**Limited Directory**—*sharer list is divided into i entries, each of which is a pointer to a cache*

Figure 8.8: Representing directory state for a block in a system with C nodes.

8.5.2 LIMITED POINTER DIRECTORY

In a chip with C caches, a complete sharer list requires C entries, one bit each, for a total of C bits. However, studies have shown that many blocks have zero sharers or one sharer. A limited pointer directory exploits this observation by having i ($i<C$) entries, where each entry requires log_2C bits, for a total of $i*log_2C$ bits, as illustrated in the bottom entry of Figure 8.8. A limited pointer directory requires some additional mechanism to handle (hopefully uncommon) situations in which the system attempts to add an $i+1^{th}$ sharer. There are three well-studied options for handling these situations, denoted using the notation Dir_iX [2, 8], where i refers to the number of pointers to sharers, and X refers to the mechanism for handling situations in which the system attempts to add an $i+1^{th}$ sharer.

- Broadcast (Dir_iB): If there are already i sharers and another GetS arrives, the directory controller sets the block's state to indicate that a subsequent GetM requires the directory to broadcast the Invalidation to all caches (i.e., a new "too many sharers" state). A drawback of Dir_iB is that the directory could have to broadcast to all C caches even when there are only K sharers ($i<K<C$), requiring the directory controller to send (and the cache controllers to process) $C-K$ unnecessary Invalidation messages. The limiting case, Dir_0B, takes this approach to the extreme by eliminating all pointers and requiring a broadcast on all coherence operations. The original Dir_0B proposal maintained two state bits per block, encoding the three MSI states plus a special "Single Sharer" state [3]. This new state helps eliminate a broadcast when a cache tries to upgrade its S copy to an M copy (similar to the

Exclusive state optimization). Similarly, the directory's I state eliminates broadcast when memory owns the block. AMD's Coherent HyperTransport [6] implements a version of Dir_0B that uses no directory state, forgoing these optimizations but eliminating the need to store any directory state. All requests sent to the directory are then broadcast to all caches.

- No Broadcast (Dir_iNB): If there are already i sharers and another GetS arrives, the directory asks one of the current sharers to invalidate itself to make room in the sharer list for the new requestor. This solution can incur significant performance penalties for widely shared blocks (i.e., blocks shared by more than i nodes), due to the time spent invalidating sharers. Dir_iNB is especially problematic for systems with coherent instruction caches because code is frequently widely shared.

- Software (Dir_iSW): If there are already i sharers and another GetS arrives, the system traps to a software handler. Trapping to software enables great flexibility, such as maintaining a full sharer list in software-managed data structures. However, because trapping to software incurs significant performance costs and implementation complexities, this approach has seen limited commercial acceptance.

8.6   DIRECTORY ORGANIZATION

Logically, the directory contains a single entry for every block of memory. Many traditional directory-based systems, in which the directory controller was integrated with the memory controller, directly implemented this logical abstraction by augmenting memory to hold the directory. For example, the SGI Origin added additional DRAM chips to store the complete directory state with each block of memory [10].

However, with today's multicore processors and large LLCs, the traditional directory design makes little sense. First, architects do not want the latency and power overhead of a directory access to off-chip memory, especially for data cached on chip. Second, system designers balk at the large off-chip directory state when almost all memory blocks are not cached at any given time. These drawbacks motivate architects to optimize the common case by caching only a subset of directory entries on chip. In the rest of this section, we discuss directory cache designs, several of which were previously categorized by Marty and Hill [13].

Like conventional instruction and data caches, a directory cache [7] provides faster access to a subset of the complete directory state. Because directories summarize the states of coherent caches, they exhibit locality similar to instruction and data accesses, but need only store each block's coherence state rather than its data. Thus, relatively small directory caches achieve high hit rates. Directory caching has no impact on the functionality of the coherence protocol; it simply reduces the average directory access latency. Directory caching has become even more important in the era of multicore processors. In older systems in which cores resided on separate chips and/or boards, message latencies were sufficiently long that they tended to amortize the directory access latency. Within a multicore processor, messages can travel from one core to another in a handful of cycles, and the latency of an off-chip directory access tends to dwarf communication latencies and become a bottleneck. Thus, for multicore processors, there is a strong incentive to implement an on-chip directory cache to avoid costly off-chip accesses.

The on-chip directory cache contains a subset of the complete set of directory entries. Thus, the key design issue is handling directory cache misses, i.e., when a coherence request arrives for a block whose directory entry is not in the directory cache.

We summarize the design options in Table 8.7 and describe them next.

Table 8.7: Comparing directory cache designs

| | Directory Cache Backed by DRAM Directory (Section 8.6.1) | Inclusive Directory Caches (Section 8.6.2) | | | | Null Directory Cache (Section 8.6.3) |
| | | Inclusive Directory Cache Embedded in Inclusive LLC (Section 8.6.2) | | Standalone Inclusive Directory Cache (Section 8.6.2) | | |
		No Recalls	With Recalls	No Recalls	With Recalls	
Directory location	DRAM	LLC		LLC		None
Uses DRAM	Yes	No		No		No
Miss at directory implies	Must access DRAM	Block must be I		Block must be I		Block could be in any state→ must broadcast
Inclusion requirements	None	LLC includes L1s		Directory cache includes L1s		None
Implementation costs	DRAM plus separate on-chip cache	Larger LLC blocks; highly associative LLC	Larger LLC blocks	Highly associative storage for redundant tags	Storage for redundant tags	None
Replacement notification	None	None	Desirable	Required	Desirable	None

8.6.1   DIRECTORY CACHE BACKED BY DRAM

The most straightforward design is to keep the complete directory in DRAM, as in traditional multi-chip multiprocessors, and use a separate directory cache structure to reduce the average access latency. A coherence request that misses in this directory cache leads to an access of this DRAM directory. This design, while straightforward, suffers from several important drawbacks. First, it requires a significant amount of DRAM to hold the directory, including state for the vast majority of blocks that are not currently cached on the chip. Second, because the directory cache is decoupled from the LLC, it is possible to hit in the LLC but miss in the directory cache, thus incurring a DRAM access even though the data is available locally. Finally, directory cache replacements must write the directory entries back to DRAM, incurring high latency and power overheads.

8.6.2   INCLUSIVE DIRECTORY CACHES

We can design directory caches that are more cost-effective by exploiting the observation that we need only cache directory states for blocks that are being cached on the chip. We refer to a directory cache as an *inclusive directory cache* if it holds directory entries for a superset of all blocks cached on the chip. An inclusive directory cache serves as a "perfect" directory cache that never misses for accesses to blocks cached on chip. There is no need to store a complete directory in DRAM. A miss in an inclusive directory cache indicates that the block is in state I; a miss is *not* the precursor to accessing some backing directory store.

We now discuss two inclusive directory cache designs, plus an optimization that applies to both designs.

Inclusive Directory Cache Embedded in Inclusive LLC

The simplest directory cache design relies on an LLC that maintains *inclusion* with the upper-level caches. Cache inclusion means that if a block is in an upper-level cache then it must also be present in a lower-level cache. For the system model of Figure 8.1, LLC inclusion means that if a block is in a core's L1 cache, then it must also be in the LLC.

A consequence of LLC inclusion is that if a block is *not* in the LLC, it is also *not* in an L1 cache and thus must be in state I for all caches on the chip. An inclusive directory cache exploits this property by embedding the coherence state of each block in the LLC. If a coherence request is sent to the LLC/directory controller and the requested address is not present in the LLC, then the directory controller knows that the requested block is not cached on-chip and thus is in state I in all the L1s.

Because the directory mirrors the contents of the LLC, the entire directory cache may be embedded in the LLC simply by adding extra bits to each block in the LLC. These added bits can lead to non-trivial overhead, depending on the number of cores and the format in which directory state is represented. We illustrate the addition of this directory state to an LLC cache block in Figure 8.9, comparing it to an LLC block in a system without the LLC-embedded directory cache.

Unfortunately, LLC inclusion has several important drawbacks. First, while LLC inclusion can be maintained automatically for private cache hierarchies (if the lower-level cache has sufficient associativity [4]), for the shared caches in our system model, it is generally necessary to send special "Recall" requests to invalidate blocks from the L1 caches when replacing a block in the LLC (discussed later in this section). More importantly, LLC inclusion requires maintaining redundant copies of cache blocks that are in upper-level caches. In multicore processors, the collective capacity of the upper-level caches may be a significant fraction of (or sometimes, even larger than) the capacity of the LLC.

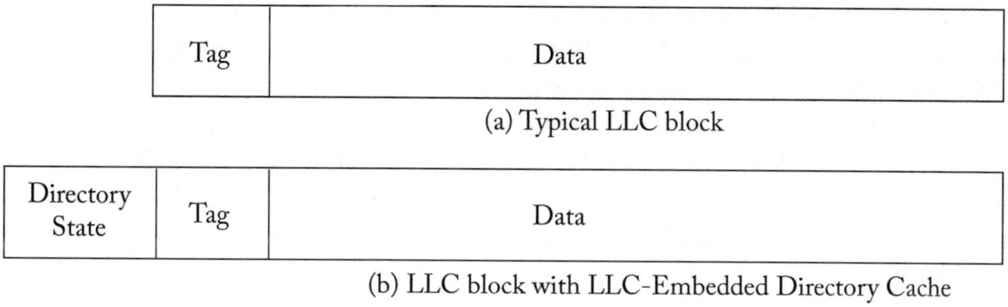

Tag	Data

(a) Typical LLC block

Directory State	Tag	Data

(b) LLC block with LLC-Embedded Directory Cache

Figure 8.9: The cost of implementing the LLC-embedded directory cache.

Standalone Inclusive Directory Cache

We now present an inclusive directory cache design that does not rely on LLC inclusion. In this design, the directory cache is a standalone structure that is logically associated with the directory controller, instead of being embedded in the LLC itself. For the directory cache to be inclusive, it must contain directory entries for the union of the blocks in all the L1 caches because a block in the LLC but not in any L1 cache must be in state I. Thus, in this design, the directory cache consists of duplicate copies of the tags at all L1 caches. Compared to the previous design, this design is more flexible, by virtue of not requiring LLC inclusion, but it has the added storage cost for the duplicate tags.

This inclusive directory cache has some significant implementation costs. Most notably, it requires a highly associative directory cache. (If we embed the directory cache in an inclusive LLC, then the LLC must also be highly associative.) Consider the case of a chip with C cores, each of which has a K-way set-associative L1 cache. The directory cache must be C*K-way associative to hold all L1 cache tags, and the associativity unfortunately grows linearly with core count. We illustrate this issue for K=2 in Figure 8.10.

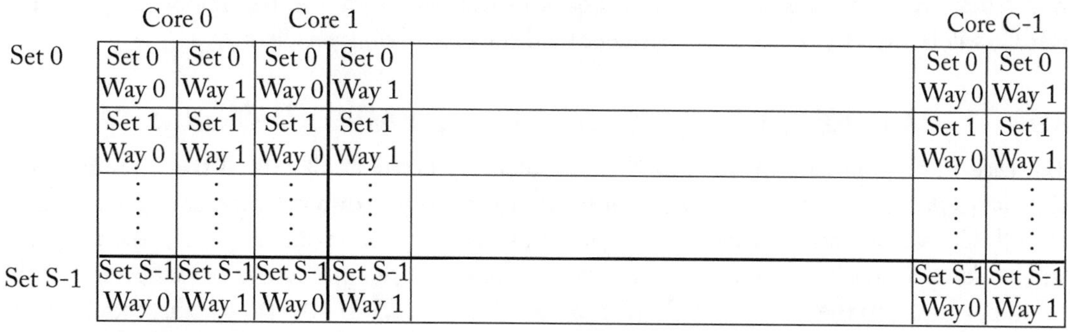

Figure 8.10: Inclusive directory cache structure (assumes 2-way L1 caches). Each entry is the tag corresponding to that set and way for the core at the top of the column.

The inclusive directory cache design also introduces some complexity, in order to keep the directory cache up-to-date. When a block is evicted from an L1 cache, the cache controller must notify the directory cache regarding which block was replaced by issuing an explicit PutS request (e.g., we cannot use a protocol with a silent eviction, as discussed in Section 8.7.4). One common optimization is to piggy-back the explicit PutS on the GetS or GetX request. Since the index bits must be the same, the PutS can be encoded by specifying which way was replaced. This is sometimes referred to as a "replacement hint," although in general it is required (and not truly a "hint").

Limiting the Associativity of Inclusive Directory Caches

To overcome the cost of the highly associative directory cache in the previous implementation, we present a technique for limiting its associativity. Rather than design the directory cache for the worst-case situation (C*K associativity), we limit the associativity by not permitting the worst-case to occur. That is, we design the directory cache to be A-way set associative, where A<C*K, and we do not permit more than A entries that map to a given directory cache set to be cached on chip. When a cache controller issues a coherence request to add a block to its cache, and the corresponding set in the directory cache is already full of valid entries, then the directory controller first evicts one of the blocks in this set from all caches. The directory controller performs this eviction by issuing a "Recall" request to all of the caches that hold this block in a valid state, and the caches respond with acknowledgments. Once an entry in the directory cache has been freed up via this Recall, then the directory controller can process the original coherence request that triggered the Recall.

The use of Recalls overcomes the need for high associativity in the directory cache but, without careful design, it could lead to poor performance. If the directory cache is too small, then Recalls will be frequent and performance will suffer. Conway et al. [6] propose a rule of thumb that the directory cache should cover at least the size of the aggregate caches it includes, but it can also be larger to reduce the rates of recalls. Also, to avoid unnecessary Recalls, this scheme works best with non-silent evictions of blocks in state S. With silent evictions, unnecessary Recalls will be sent to caches that no longer hold the block being recalled.

8.6.3 NULL DIRECTORY CACHE (WITH NO BACKING STORE)

The least costly directory cache is to have no directory cache at all. Recall that the directory state helps prune the set of coherence controllers to which to forward a coherence request. But as with Coarse Directories (Section 8.5.1), if this pruning is done incompletely, the protocol still works correctly, but unnecessary messages are sent and the protocol is less efficient than it could be. Taken to the extreme, a Dir_0B protocol (Section 8.5.2) does no pruning whatsoever, in which case it does not actually need a directory at all (or a directory cache). Whenever a coherence request arrives at the directory controller, the directory controller simply forwards it to all caches (i.e., broadcasts the forwarded request). This directory cache design, which we call the Null Directory Cache, may seem simplistic, but it is popular for small-to medium-scale systems because it incurs no storage cost.

One might question the purpose of a directory controller if there is no directory state, but it serves two important roles. First, as with all other systems in this chapter, the directory controller is responsible for the LLC; it is, more precisely, an LLC/directory controller. Second, the directory controller serves as an ordering point in the protocol; if multiple cores concurrently request the same block, the requests are ordered at the directory controller. The directory controller resolves which request happens first.

8.7   PERFORMANCE AND SCALABILITY OPTIMIZATIONS

In this section, we discuss several optimizations to improve the performance and scalability of directory protocols.

8.7.1   DISTRIBUTED DIRECTORIES

So far we have assumed that there is a single directory attached to a single monolithic LLC. This design clearly has the potential to create a performance bottleneck at this shared, central resource. The typical, general solution to the problem of a centralized bottleneck is to distribute the resource. The directory for a given block is still fixed in one place, but different blocks can have different directories.

In older, multi-chip multiprocessors with N *nodes*—each node consisting of multiple chips, including the processor core and memory—each node typically had 1/N of the memory associated with it and the corresponding 1/Nth of the directory state.

We illustrate such a system model in Figure 8.11. The allocation of memory addresses to nodes is often static and often easily computable using simple arithmetic. For example, in a system with N directories, block B's directory entry might be at directory *B modulo N*. Each block has a *home*, which is the directory that holds its memory and directory state. Thus, we end up with a system in which there are multiple, independent directories managing the coherence for different sets of blocks. Having multiple directories provides greater bandwidth of coherence transactions than requiring all coherence traffic to pass through a single, central resource. Importantly, distributing the directory has no impact on the coherence protocol.

In today's world of multicore processors with large LLCs and directory caches, the approach of distributing the directory is logically the same as in the traditional multi-chip multiprocessors. We can distribute (bank) the LLC and directory cache. Each block has a home bank of the LLC with its associated bank of the directory cache.

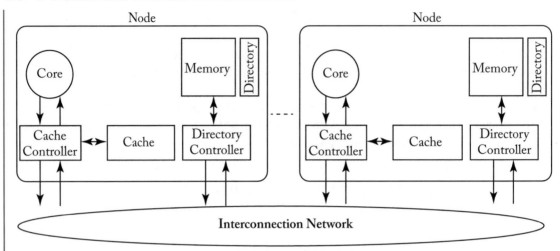

Figure 8.11: Multiprocessor system model with distributed directory.

8.7.2 NON-STALLING DIRECTORY PROTOCOLS

One performance limitation of the protocols presented thus far is that the coherence controllers stall in several situations. In particular, the cache controllers stall when they receive forwarded requests for blocks in certain transient states, such as IM^A. In Tables 8.8 and 8.9, we present a variant of the baseline MSI protocol that does not stall in these scenarios. For example, when a cache controller has a block in state IM^A and receives a Fwd-GetS, it processes the request and changes the block's state to IM^AS. This state indicates that after the cache controller's GetM transaction completes (i.e., when the last Inv-Ack arrives), the cache controller will change the block state to S. At this point, the cache controller must also send the block to the requestor of the GetS and to the directory, which is now the owner. By not stalling on the Fwd-GetS, the cache controller can improve performance by continuing to process other forwarded requests behind that Fwd-GetS in its incoming queue.

A complication in the non-stalling protocol is that, while in state IM^AS, a Fwd-GetM could arrive. Instead of stalling, the cache controller processes this request and changes the block's state to IM^ASI (in I, going to M, waiting for Inv-Acks, then going to S and then to I). A similar set of transient states arises for blocks in SM^A. Removing stalling leads to more transient states, in general, because the coherence controller must track (using new transient states) the additional messages it is processing instead of stalling.

We did not remove the stalls from the directory controller. As with the memory controllers in the snooping protocols in Chapter 7, we would need to add an impractically large number of transient states to avoid stalling in all possible scenarios.

Table 8.8: Non-stalling MSI directory protocol—cache controller

	Load	Store	Replacement	Fwd-GetS	Fwd-GetM	Inv	Put-Ack	Data from Dir (ack=0)	Data from Dir (ack>0)	Data from Owner	Inv-Ack	Last-Inv-Ack
I	Send GetS to Dir/IS^D	Send GetM to Dir/IM^{AD}										
IS^D	Stall	Stall	Stall			Send Inv-Ack to Req /IS^{D}I		-/S		-/S		
IS^{D}I	**Stall**	**Stall**	**Stall**					-/I		-/I		
IM^{AD}	Stall	Stall	Stall	Stall	Stall			-/M	-/IM^A	-/M	ack--	
IM^A	Stall	Stall^D	Stall	-/IM^{A}S	-/IM^{A}I						ack--	-/M
IM^{A}S	**Stall**	**Stall**	**Stall**			Send Inv-Ack to Req/ IM^{A}SI					**ack--**	Send data to Req and Dir/S
IM^{A}SI	**Stall**	**Stall**	**Stall**								ack--	Send data to Req and Dir/I
IM^{A}I	**Stall**	**Stall**	**Stall**								ack--	Send data to Req/I
S	Hit	Send GetM to Dir/ SM^{AD}	Send PutS to Dir /SI^A			Send Inv-Ack to Req/I						
SM^{AD}	Hit	Stall	Stall	Stall	Stall	Send Inv-Ack to Req/ IM^{AD}		-/M	-/SM^A		ack--	
SM^A	**Hit**	**Stall**	**Stall**	-/SM^{A}S	-/SM^{A}I						ack--	-/M
SM^{A}S	Stall	Stall	Stall			Send Inv-Ack to Req/ SM^{A}SI					ack--	Send data to Req and Dir/S
SM^{A}SI	Stall	Stall	Stall								ack--	Send data to Req and Dir/I
SM^{A}I	Stall	Stall	Stall								ack--	Send data to Req/I
M	Hit	Hit	Send PutM +data to Dir/MI^A	Send data to Req and Dir/S	Send data to Req/I							
MI^A	Stall	Stall	Stall	Send data to Req and Dir/SI^A	Send data to Req/II^A		-/I					
SI^A	Stall	Stall	Stall			Send Inv-Ack to Req/ II^A	-/I					
II^A	Stall	Stall	Stall				-/I					

Table 8.9: Non-stalling MSI directory protocol—directory controller

	GetS	GetM from Owner	PutS-NotLast	PutS-Last	PutM+data from Owner	PutM+data from Non-Owner	Data
I	Send data to Req, add Req to Sharer/S	Send data to Req, set Owner to Req/M	Send Put-Ack to Req	Send Put-Ack to Req		Send Put-Ack to Req	
S	Send data to Req, add Req to Sharers	Send data to Req, set Owner to Req, send Inv to Sharers, clear Sharers/M	Send Put-Ack to Req, remove Req from Sharers	Send Put-Ack to Req, remove Req from Sharers/I		Remove Req from Sharers, send Put-Ack to Req	
M	Forward GetS to Owner, add Req to Sharers, clear Owner/S^D	Forward GetM to Owner, set Owner to Req	Send Put-Ack to Req	Send Put-Ack to Req	Copy data to memory, send Put-Ack to Req, clear Owner/I	Send Put-Ack to Req	
S^D	Stall	Stall	Sent Put-Ack to Req, remove Req from Sharers	Sent Put-Ack to Req, remove Req from Sharers		Remove Req from Sharers, send Put-Ack to Req	Copy data to memory/S

8.7.3 INTERCONNECTION NETWORKS WITHOUT POINT-TO-POINT ORDERING

We mentioned in Section 8.2, when discussing the system model of our baseline MSI directory protocol, that we assumed that the interconnection network provides point-to-point ordering for the Forwarded Request network. At the time, we claimed that point-to-point ordering simplifies the architect's job in designing the protocol because ordering eliminates the possibility of certain races.

We now present one example race that is possible if we do not have point-to-point ordering in the interconnection network. We assume the MOSI protocol from Section 8.4. Core C1's cache owns a block in state M. Core C2 sends a GetS request to the directory and core C3 sends a GetM request to the directory. The directory receives C2's GetS first and then C3's GetM. For both requests, the directory forwards them to C1.

- With point-to-point ordering (illustrated in Figure 8.12): C1 receives the Fwd-GetS, responds with Data, and changes the block state to O. C1 then receives the Fwd-GetM, responds with Data, and changes the block state to I. This is the expected outcome.

- Without point-to-point ordering (illustrated in Figure 8.13): The Fwd-GetM from C3 may arrive at C1 first. C1 responds with Data to C3 and changes the block state to I. The

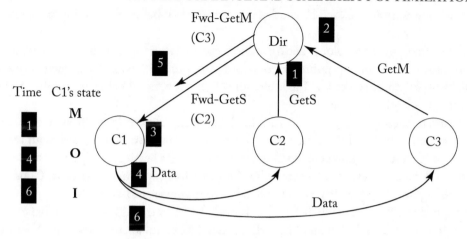

Figure 8.12: Example with point-to-point ordering.

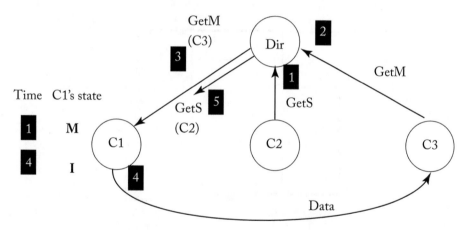

Figure 8.13: Example without point-to-point ordering. Note that C2's Fwd-GetS arrives at C1 in state I and thus C1 does not respond.

Fwd-GetS from C2 then arrives at C1. C1 is in I and cannot respond. The GetS request from C2 will never be satisfied and the system will eventually deadlock.

The directory protocols we have presented thus far are not compatible with interconnection networks that do not provide point-to-point order for the Forwarded Request network. To make the protocols compatible, we would have to modify them to correctly handle races like the one described above. One typical approach to eliminating races like these is to add extra handshaking messages. In the example above, the directory could wait for the cache controller to acknowledge reception of each forwarded request sent to it before forwarding another request to it.

Given that point-to-point ordering reduces complexity, it would seem an obvious design decision. However, enforcing point-to-point ordering prohibits us from implementing some potentially useful optimizations in the interconnection network. Notably, it prohibits the unrestricted use of adaptive routing.

Adaptive routing enables a message to dynamically choose its path as it traverses the network, generally to avoid congested links or switches. Adaptive routing, although useful for spreading traffic and mitigating congestion, enables messages between endpoints to take different paths and thus arrive in a different order than that in which they were sent. Consider the example in Figure 8.14, in which Switch A sends two messages, M1 and then M2, to Switch D. With adaptive routing, they take different paths, as shown in the figure. If there happens to be more congestion at Switch B than at Switch C, then M2 could arrive at Switch D before M1, despite being sent after M1.

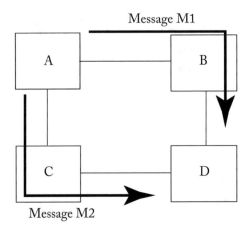

Figure 8.14: Adaptive routing example.

8.7.4 SILENT VS. NON-SILENT EVICTIONS OF BLOCKS IN STATE S

We designed our baseline directory protocol such that a cache cannot silently evict a block in state S (i.e., without issuing a PutS to notify the directory). To evict an S block, the cache must send a PutS to the directory and wait for a Put-Ack. Another option would be to allow silent evictions of S blocks. (A similar discussion could be made for blocks in state E, if one considers the E state to not be an ownership state, in which case silent evictions of E blocks are possible.)

Advantages of Silent PutS

The drawback to the explicit PutS is that it uses interconnection network bandwidth—albeit for small data-free PutS and Put-Ack messages—even in cases when it ends up not being helpful. For example if core C1 sends a PutS to the directory and then subsequently wants to perform a load to this block, C1 sends a GetS to the directory and re-acquires the block in S. If C1 sends this second GetS before any intervening GetM requests from other cores, then the PutS transaction served no purpose but did consume bandwidth.

Advantages of Explicit PutS

The primary motivation for sending a PutS is that a PutS enables the directory to remove the cache no longer sharing the block from its list of sharers. There are three benefits to having a more precise sharer list. First, when a subsequent GetM arrives, the directory need not send an Invalidation to this cache. The GetM transaction is accelerated by eliminating the Invalidation and having to wait for the subsequent Inv-Ack. Second, in a MESI protocol, if the directory is precisely counting the sharers, it can identify situations in which the last sharer has evicted its block; when the directory knows there are no sharers, it can respond to a GetS with Exclusive data. Third, recall from Section 8.6.2 that directory caches that use Recalls can benefit from having explicit PutS messages to avoid unnecessary Recall requests.

A secondary motivation for sending a PutS, and the reason our baseline protocol does send a PutS, is that it simplifies the protocol by eliminating some races. Notably, without a PutS, a cache that silently evicts a block in S and then sends a GetS request to re-obtain that evicted block in S can receive an Invalidation from the directory before receiving the data for its GetS. In this situation, the cache does not know if the Invalidation pertains to the first period in which it held the block in S or the second period (i.e., whether the Invalidation is serialized before or after the GetS). The simplest solution to this race is to pessimistically assume the worst case (the Invalidation pertains to the second period) and always invalidate the block as soon as its data arrives. More efficient solutions exist, but complicate the protocol.

8.8   CASE STUDIES

In this section, we discuss several commercial directory coherence protocols. We start with a traditional multi-chip system, the SGI Origin 2000. We then discuss more recently developed directory protocols, including AMD's Coherent HyperTransport and the subsequent HyperTransport Assist. Last, we present Intel's QuickPath Interconnect (QPI).

8.8.1   SGI ORIGIN 2000

The Silicon Graphics Origin 2000 [10] was a commercial multi-chip multiprocessor designed in the mid-1990s to scale to 1024 cores. The emphasis on scalability necessitated a scalable coherence protocol, resulting in one of the first commercial shared-memory systems using a directory protocol. The Origin's directory protocol evolved from the design of the Stanford DASH multiprocessor [11], as the DASH and Origin had overlapping architecture teams.

As illustrated in Figure 8.15, the Origin consists of up to 512 nodes, where each node consists of two MIPS R10000 processors connected via a bus to a specialized ASIC called the Hub. Unlike similar designs, Origin's processor bus does not exploit coherent snooping and simply connects the processors to each other and to the node's Hub. The Hub manages the cache coherence protocol and interfaces the node to the interconnection network. The Hub also connects to the node's portion of the distributed memory and directory. The network does not support any ordering, even point-to-point ordering between nodes. Thus, if Processor A sends two messages to Processor B, they may arrive in a different order than that in which they were sent.

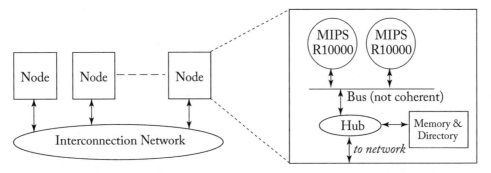

Figure 8.15: SGI Origin.

The Origin's directory protocol has a few distinguishing features that are worth discussing. First, because of its scalability, each directory entry contains fewer bits than necessary to represent every possible cache that could be sharing a block. The directory dynamically chooses, for each directory entry, to use either a coarse bit vector representation or a limited pointer representation (Section 8.5).

A second interesting feature in the protocol is that because the network provides no ordering, there are several new coherence message race conditions that are possible. Notably, the examples from Section 8.7.3 are possible. To maintain correctness, the protocol must consider all of these possible race conditions introduced by not enforcing ordering in the network.

A third interesting feature is the protocol's use of a non-ownership E state. Because the E state is not an ownership state, a cache can silently evict a block in state E (or state S).

A fourth interesting feature of the Origin is that is provides a special Upgrade coherence request to transition from S to M without needlessly requesting data, which is not unusual but does introduce a new race. There is a window of vulnerability between when processor P1 sends an Upgrade and when the Upgrade is serialized at the directory; if another processor's GetM or Upgrade is serialized first, then P1's state is I when its Upgrade arrives at the directory, and P1 in fact needs data. In this situation, the directory sends a negative acknowledgment (NACK) to P1, and P1 must send a GetM to the directory.

Another interesting feature of the Origin's E state is how requests are satisfied when a processor is in E. Consider the case where processor P1 obtains a block in state E. If P2 now sends a GetS to the directory, the directory must consider that P1 (a) might have silently evicted the block, (b) might have an unmodified value of the block (i.e., with the same value as at memory), or (c) might have a modified value of the block. To handle all of these possibilities, the directory responds with data to P2 and also forwards the request to P1. P1 sends P2 either new data (if in M) or just an acknowledgment. P2 must wait for both responses to arrive to know which message's data to use.

One other quirk of the Origin is that it uses only two networks (request and response) instead of the three required to avoid deadlock. A directory protocol has three message types (request, forwarded request, and response) and thus nominally requires three networks. Instead, the Origin protocol detects when deadlock could occur and sends a "backoff" message to a requestor on the response network. The backoff message contains the list of nodes that the request needs to be sent to, and the requestor can then send to them on the request network.

8.8.2   COHERENT HYPERTRANSPORT

Directory protocols were originally developed to meet the needs of highly scalable systems, and the SGI Origin is a classic example of such a system. Recently, however, directory protocols have become attractive even for small- to medium-scale systems because they facilitate the use of point-to-point links in the interconnection network. This advantage of directory protocols motivated the design of AMD's Coherent HyperTransport (HT) [5]. Coherent HT enables glueless connections of AMD processors into small-scale multiprocessors. Perhaps ironically, Coherent HT actually uses broadcasts, thus demonstrating that the appeal of directory protocols in this case is the use of point-to-point links, rather than scalability.

AMD observed that systems with up to eight processor chips can be built with only three point-to-point links per chip and a maximum chip-to-chip distance of three links. Eight processor chips, each with 6 cores, means a system with a respectable 48 cores. To keep the protocol simple, Coherent HT uses a variation on a Dir_0B directory protocol (Section 8.5.2) that stores no stable directory state. Any coherence request sent to the directory is forwarded to all cache controllers (i.e., broadcast). Coherent HT can also be thought of as an example of a null directory cache: requests always miss in the (null) directory cache, so it always broadcasts. Because of the broadcasts, the protocol does not scale to large-scale systems, but that was not the goal.

In a system with Coherent HT, each processor chip contains some number of cores, one or more integrated memory controllers, one or more integrated HyperTransport controllers, and between one and three Coherent HT links to other processor chips. A "node" consists of a processor chip and its associated memory for which it is the home.

There are many viable interconnection network topologies, such as the four-node system shown in Figure 8.16. Significantly, this protocol does not require a total order of coherence requests, which provides greater flexibility for the interconnection network.

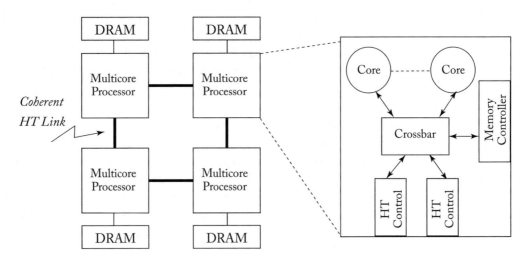

Figure 8.16: Four-node coherent HyperTransport system (adapted from [5]).

A coherence transaction works as follows. A core unicasts a coherence request to the directory controller at the home node, as in a typical directory protocol. Because the directory has no state and thus cannot determine which cores need to observe the request, the directory controller then broadcasts the forwarded request to all cores, including the requestor. (This broadcast is like what happens in a snooping protocol, except that the broadcast is not totally ordered and does not originate with the requestor.) Each core then receives the forwarded request and sends a response (either data or an acknowledgment) to the requestor. Once the requestor has received all of the responses, it sends a message to the directory controller at the home node to complete the transaction.

Looking at this protocol, one can view it as the best or worst of both worlds. Optimistically, it has point-to-point links with no directory state or complexity, and it is sufficiently scalable for up to eight processors. Pessimistically, it has the long three-hop latency of directories—or four hops, if you consider the message from the requestor to the home to complete the transaction, although this message is not on the critical path—with the high broadcast traffic of snooping. In fact, Coherent HT uses even more bandwidth than snooping because all broadcasted forwarded requests generate a response. The drawbacks of Coherent HT motivated an enhanced design called HyperTransport Assist [6].

8.8.3   HYPERTRANSPORT ASSIST

For the 12-core AMD Opteron processor-based system code-named Magny Cours, AMD developed HyperTransport Assist [6]. HT Assist enhances Coherent HT by eliminating the need to broadcast every forwarded coherence request. Instead of the Dir_0B-like protocol of Coherent HT, HT Assist uses a directory cache similar to the design described in Section 8.6.2. Each multicore processor chip has an inclusive directory cache that has a directory entry for every block (a) for which it is the home and (b) that is cached anywhere in the system. There is no DRAM directory, thus preserving one of the key features of Coherent HT. A miss in the directory cache indicates that the block is not cached anywhere. HT Assist's directory cache uses Recall requests to handle situations in which the directory cache is full and needs to add a new entry. Although HT Assist appears from our description thus far to be quite similar to the design in Section 8.6.2, it has several distinguishing features that we present in greater detail.

First, the directory entries provide only enough information to determine whether a coherence request must be forwarded to all cores, forwarded to a single core, or satisfied by the home node's memory. That is, the directory does not maintain sufficient state to distinguish needing to forward a request to two cores from needing to forward the request to all cores. This design decision eliminated the storage cost of having to maintain the exact number of sharers of each block; instead, two directory states distinguish "one sharer" from "more than one sharer."

Second, the HT Assist design is careful to avoid incurring a large number of Recalls. AMD adhered to a rule of thumb that there should be at least twice as many directory entries as cached blocks. Interestingly, AMD chose not to send explicit PutS requests; their experiments apparently convinced them that the additional PutS traffic was not worth the limited benefit in terms of a reduction in Recalls.

Third, the directory cache shares the LLC. The LLC is statically partitioned by the BIOS at boot time, and the default is to allocate 1 MB to the directory cache and allocate the remaining 5 MB to the LLC itself. Each 64-byte block of the LLC that is allocated to the directory cache is interpreted as 16 4-byte directory entries, organized as four 4-way set-associative sets.

8.8.4   INTEL QPI

Intel developed its QuickPath Interconnect (QPI) [9, 12] for connecting processor chips starting with the 2008 Intel Core microarchitecture, and QPI first shipped in the Intel Core i7-9xx processor. Prior to this, Intel connected processor chips with a shared-wire bus called the Front-Side Bus (FSB). FSB evolved from a single shared bus to multiple buses, but the FSB approach was fundamentally bottlenecked by the electrical signaling limits of the buses. To overcome this limitation, Intel designed QPI to connect processor chips with point-to-point (i.e., non-bus) links. QPI specifies multiple levels of the networking stack, from physical layer to protocol layer. For purposes of this primer, we focus on the protocol layer here.

QPI supports five stable coherence states, the typical MESI states and the F(orward) state. The F state is a clean, read-only state, and it is distinguished from the S state because a cache with a block in F may respond with data (i.e., forward the data) to coherence requests. Only one cache may hold a block in F at a given time. The F state is somewhat similar to the O state, but differs in that a block in F is not dirty and can thus be silently evicted; a cache that wishes to evict a block in O must copy the block back to memory. The benefit of the F state is that it allows read-only data to be sourced from a cache, which is often faster than sourcing it from memory (which usually responds to requests when a block is read-only).

QPI provides two different protocol modes, depending on the size of the system: "home snoop" and "source snoop."

QPI's Home Snoop mode is effectively a scalable directory protocol (i.e., do not be confused by the word "snoop" in its name[2]). As with typical directory protocols, a core C1 issues a request to the directory at the home node C2, and the directory forwards that request to only the node(s) that need to see it, say C3 (the owner in M). C3 responds with data to C1 and also sends a message to C2 to notify the directory. When the directory at C2 receives the notification from C3, it sends a "completion" message to C1, at which point C1 may use the data it received from C3. The directory serves as the serialization point in the protocol and resolves message races.

QPI's Source Snoop protocol mode is designed to have lower-latency coherence transactions, at the expense of not scaling well to large systems with many nodes. A core C1 broadcasts a request to all nodes, including the home. Each core responds to the home with a "snoop response" that indicates what state the block was in at that core; if the block was in state M, then the core sends the block to the requestor in addition to the snoop response to the home. Once the home has received all of the snoop responses for a request, the request has been ordered. At this point, the home either sends data to the requestor (if no core owned the block in M) or a non-data message to the requestor; either message, when received by the requestor, completes the transaction.

Source Snoop's use of broadcast requests is similar to a snooping protocol, but with the critical difference of the broadcast requests not traveling on a totally ordered broadcast network. Because the network is not totally ordered, the protocol must have a mechanism to resolve races (i.e., when two broadcasts race, such that core C1 sees broadcast A before broadcast B and core C2 sees B before A). This mechanism is provided by the home node, albeit in a way that differs from typical race ordering in directory protocols. Typically, the directory at the requested block's home orders two racing requests based on which request arrives at the home first. QPI's Source Snoop, instead, orders the requests based on which request's snoop responses have all arrived at the home.

Consider the race situation in which block A is initially in state I in the caches of both C1 and C2. C1 and C2 both decide to broadcast GetM requests for A (i.e., send a GetM to the other core and to the home). When each core receives the other core's GetM, it sends a snoop response to the home. Assume that C2's snoop response arrives at the home before C1's

---

[2]Intel uses the word "snoop" to refer to what a core does when it receives a coherence request from another node.

snoop response. In this case, C1's request is ordered first and the home sends data to C1 and informs C1 that there is a race. C1 then sends an acknowledgment to the home, and the home subsequently sends a message to C1 that both completes C1's transaction and tells C1 to send the block to C2. Handling this race is somewhat more complicated than in a typical directory protocol in which requests are ordered when they arrive at the directory.

Source Snoop mode uses more bandwidth than Home Snoop, due to broadcasting, but Source Snoop's common case (no race) transaction latency is less. Source Snoop is somewhat similar to Coherent HyperTransport, but with one key difference. In Coherent HT, a request is unicasted to the home, and the home broadcasts the request. In Source Snoop, the requestor broadcasts the request. Source Snoop thus introduces more complexity in resolving races because there is no single point at which requests can be ordered; Coherent HT uses the home for this purpose.

8.9   DISCUSSION AND THE FUTURE OF DIRECTORY PROTOCOLS

Directory protocols have come to dominate the market. Even in small-scale systems, directory protocols are more common that snooping protocols, largely because they facilitate the use of point-to-point links in the interconnection network. Furthermore, directory protocols are the *only* option for systems requiring scalable cache coherence. Although there are numerous optimizations and implementation tricks that can mitigate the bottlenecks of snooping, fundamentally none of them can eliminate these bottlenecks. For systems that need to scale to hundreds or even thousands of nodes, a directory protocol is the only viable option for coherence. Because of their scalability, we anticipate that directory protocols will continue their dominance for the foreseeable future.

It is possible, though, that future highly scalable systems will not be coherent or at least not coherent across the entire system. Perhaps such systems will be partitioned into subsystems that are coherent, but coherence is not maintained across the subsystems. Or perhaps such systems will follow the lead of supercomputers, like those from Cray, that have either not provided coherence [14] or have provided coherence but restricted what data can be cached [1].

8.10   REFERENCES

[1] D. Abts, S. Scott, and D. J. Lilja. So many states, so little time: Verifying memory coherence in the Cray X1. In *Proc. of the International Parallel and Distributed Processing Symposium*, 2003. DOI: 10.1109/ipdps.2003.1213087. 189

[2] A. Agarwal, R. Simoni, M. Horowitz, and J. Hennessy. An evaluation of directory schemes for cache coherence. In *Proc. of the 15th Annual International Symposium on Computer Architecture*, pp. 280–89, May 1988. DOI: 10.1109/isca.1988.5238. 171

[3] J. K. Archibald and J.-L. Baer. An economical solution to the cache coherence problem. In *Proc. of the 11th Annual International Symposium on Computer Architecture*, pp. 355–62, June 1984. DOI: 10.1145/800015.808205. 171

[4] J.-L. Baer and W.-H. Wang. On the inclusion properties for multi-level cache hierarchies. In *Proc. of the 15th Annual International Symposium on Computer Architecture*, pp. 73–80, May 1988. DOI: 10.1109/isca.1988.5212. 174

[5] P. Conway and B. Hughes. The AMD Opteron northbridge architecture. *IEEE Micro*, 27(2):10–21, March/April 2007. DOI: 10.1109/mm.2007.43. 185, 186

[6] P. Conway, N. Kalyanasundharam, G. Donley, K. Lepak, and B. Hughes. Cache hierarchy and memory subsystem of the AMD Opteron processor. *IEEE Micro*, 30(2):16–29, March/April 2010. DOI: 10.1109/mm.2010.31. 172, 176, 186, 187

[7] A. Gupta, W.-D. Weber, and T. Mowry. Reducing memory and traffic requirements for scalable directory-based cache coherence schemes. In *Proc. of the International Conference on Parallel Processing*, 1990. DOI: 10.1007/978-1-4615-3604-8_9. 172

[8] M. D. Hill, J. R. Larus, S. K. Reinhardt, and D. A. Wood. Cooperative shared memory: Software and hardware for scalable multiprocessors. *ACM Transactions on Computer Systems*, 11(4):300–18, November 1993. DOI: 10.1145/161541.161544. 171

[9] Intel Corporation. An Introduction to the Intel QuickPath Interconnect. Document Number 320412–001US, January 2009. 187

[10] J. Laudon and D. Lenoski. The SGI Origin: A ccNUMA highly scalable server. In *Proc. of the 24th Annual International Symposium on Computer Architecture*, pp. 241–51, June 1997. DOI: 10.1145/264107.264206. 162, 172, 183

[11] D. Lenoski, J. Laudon, K. Gharachorloo, W.-D. Weber, A. Gupta, J. Hennessy, M. Horowitz, and M. Lam. The Stanford DASH multiprocessor. *IEEE Computer*, 25(3):63–79, March 1992. DOI: 10.1109/2.121510. 184

[12] R. A. Maddox, G. Singh, and R. J. Safranek. *Weaving High Performance Multiprocessor Fabric: Architecture Insights into the Intel QuickPath Interconnect*. Intel Press, 2009. 187

[13] M. R. Marty and M. D. Hill. Virtual hierarchies to support server consolidation. In *Proc. of the 34th Annual International Symposium on Computer Architecture*, June 2007. DOI: 10.1145/1250662.1250670. 172

[14] S. L. Scott. Synchronization and communication in the Cray T3E multiprocessor. In *Proc. of the 7th International Conference on Architectural Support for Programming Languages and Operating Systems*, pp. 26–36, October 1996. 189

CHAPTER 9

Advanced Topics in Coherence

In Chapters 7 and 8, we have presented snooping and directory coherence protocols in the context of the simplest system models that were sufficient for explaining the fundamental issues of these protocols. In this chapter, we extend our presentation of coherence in several directions. In Section 9.1, we discuss the issues involved in designing coherence protocols for more sophisticated system models. In Section 9.2, we describe optimizations that apply to both snooping and directory protocols. In Section 9.3, we explain how to ensure that a coherence protocol remains live (i.e., avoids deadlock, livelock, and starvation). In Section 9.4, we present token coherence protocols [12], a class of protocols that subsumes both snooping and directory protocols. We conclude in Section 9.5 with a brief discussion of the future of coherence.

9.1 SYSTEM MODELS

Thus far, we have assumed a simple system model, in which each processor core has a single-level write-back data cache that is physically addressed. This system model omitted numerous features that are typically present in commercial systems, such as instruction caches (Section 9.1.1), translation lookaside buffers (Section 9.1.2), virtually addressed caches (Section 9.1.3), write-through caches (Section 9.1.4), coherent DMA (Section 9.1.5), and multiple levels of caches (Section 9.1.6).

9.1.1 INSTRUCTION CACHES

All modern cores have at least one level of instruction cache, raising the question of whether and how to support instruction cache coherence. Although truly self-modifying code is rare, cache blocks containing instructions may be modified when the operating system loads a program or library, a just-in-time (JIT) compiler generates code, or a dynamic run-time system re-optimizes a program.

Adding instruction caches to a coherence protocol is superficially straightforward; blocks in an instruction cache are read-only and thus in either stable state I or S. Furthermore, the core never writes directly to the instruction cache; a core modifies code by performing stores to its data cache. Thus, the instruction cache's coherence controller takes action only when it observes a GetM from another cache (possibly its own L1 data cache) to a block in state S and simply invalidates the block.

Instruction cache coherence differs from data cache coherence for several reasons. Most importantly, once fetched, an instruction may remain buffered in the core's pipeline for many cycles (e.g., consider a core that fills its 128-instruction window with a long sequence of loads, each of which misses all the way to DRAM). Software that modifies code needs some way to know when a write has affected the fetched instruction stream. Some architectures, such as the AMD Opteron, address this issue using a separate structure that tracks which instructions are in the pipeline. If this structure detects a change to an in-flight instruction, it flushes the pipeline. However, because instructions are modified far less frequently than data, other architectures require the software to explicitly manage coherence. For example, the Power architecture provides the icbi (instruction cache block invalidate) instruction to invalidate an instruction cache entry.

9.1.2   TRANSLATION LOOKASIDE BUFFERS (TLBs)

Translation lookaside buffers (TLBs) are caches that hold a special type of data: translations from virtual to physical addresses. As with other caches, they must be kept coherent. Like instruction caches, they have not historically participated in the same all-hardware coherence protocols that handle data caches. The traditional approach to TLB coherence is *TLB shootdown* [18], a software-managed coherence scheme that may or may not have some hardware support. In a classic implementation, a core invalidates a translation entry (e.g., by clearing the page table entry's PageValid bit) and sends an inter-processor interrupt to all cores. Each core receives its interrupt, traps to a software handler, either invalidates the specific translation entry from its TLBs or flushes all entries from its TLBs (depending upon the platform). Each core must also ensure that there are no instructions in flight which are using the now-stale translation, typically by flushing the pipeline. Each core then sends an acknowledgment back to the initiating core, using an interprocessor interrupt. The initiating core waits for all of the acknowledgments, ensuring that all stale translation entries have been invalidated, before modifying the translation (or reusing the physical page). Some architectures provide special support to accelerate TLB shootdown. For example, the Power architecture eliminates costly inter-processor interrupts by using a special tlbie (TLB invalidate entry) instruction; the initiating core executes a tlbie instruction, which broadcasts the invalidated virtual page number to all the cores and completes only once all cores have completed the invalidation.

Recent research proposed eliminating TLB shootdown and instead incorporating the TLBs into the existing all-hardware coherence protocol for the data and instruction caches [16]. This all-hardware solution is more scalable than TLB shootdown, but it requires a modification to the TLBs to enable them to be addressable in the same way as the data and instruction caches. That is, the TLBs must snoop the physical addresses of the blocks that hold translations in memory.

9.1.3   VIRTUAL CACHES

Most caches in current systems—and all caches discussed thus far in this primer—are accessed with physical addresses, yet caches can also be accessed with virtual addresses. We illustrate both options in Figure 9.1. A virtually addressed cache ("virtual cache") has one key advantage with

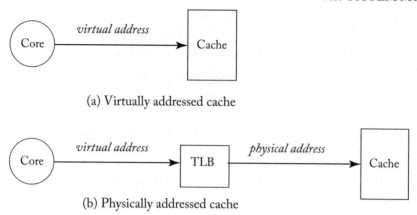

(a) Virtually addressed cache

(b) Physically addressed cache

Figure 9.1: Physical vs. virtual addressed caches.

respect to a physically addressed cache ("physical cache"): the latency of address translation is off the critical path.[1] This latency advantage is appealing for level-one caches, where latency is critical, but generally less compelling for lower level caches where latencies are less critical. Virtual caches, however, pose a few challenges to the architect of a coherence protocol:

- Coherence protocols invariably operate on physical addresses, for compatibility with main memory, which would otherwise require its own TLB. Thus, when a coherence request arrives at a virtual cache, the request's address must undergo reverse translation to obtain the virtual address with which to access the cache.

- Virtual caches introduce the problem of *synonyms*. Synonyms are multiple virtual addresses that map to the same physical address. Without mechanisms in place to avoid synonyms, it is possible for synonyms to simultaneously exist in a virtual cache. Thus, not only does a virtual cache requires a mechanism for reverse translation but also any given reverse translation could result in multiple virtual addresses.

Because of the complexity of implementing virtual caches, they are rarely used in current systems. However, they have been used in a number of earlier systems, and it is possible that they could become more relevant again in the future.

9.1.4 WRITE-THROUGH CACHES

Our baseline system model assumes writeback L1 data caches and a shared writeback LLC. The other option, write-through caches, has several advantages and disadvantages. The obvious disadvantages include significantly greater bandwidth and power to write data through to the

---

[1]A cache that is virtually indexed and physically tagged has this same advantage without the shortcomings of virtual caches.

next lower level of the memory hierarchy. In modern systems, these disadvantages effectively limit the write-through/writeback decision to the L1 cache.

The advantages of write-through L1s include the following.

- A significantly simpler two-state VI (Valid and Invalid) coherence protocol. Stores write through to the LLC and invalidate all Valid copies in other caches.

- An L1 eviction requires no action, besides changing the L1 state to Invalid, because the LLC always holds up-to-date data.

- When the LLC handles a coherence request, it can respond immediately because it always has up-to-date data.

- When an L1 observes another core's write, it needs only to change the cache block's state to Invalid. Importantly, this allows the L1 to represent each block's state with a single, clearable flip-flop, eliminating complex arbitration or dual-ported state RAMs.

- Finally, write-through caches also facilitate fault tolerance. Although a detailed discussion is outside the scope of this primer, a write-through L1 cache never holds the only copy of a block because the LLC always holds a valid copy. This allows the L1 to use only parity because it can always just invalidate a block with a parity error.

Write-through caches pose some challenges with multithreaded cores and shared L1 caches. Recall that TSO requires write atomicity, and thus all threads (except the thread performing the store) must see the store at the same time. Thus, if two threads T0 and T1 share the same L1 data cache, T0's store to block A must prevent T1 from accessing the new value until all copies in other caches have been invalidated (or updated). Despite these complications and disadvantages, several designs use write-through L1 caches, including the Sun Niagara processors and AMD Bulldozer.

9.1.5   COHERENT DIRECT MEMORY ACCESS (DMA)

In Chapter 2, when we first introduced coherence, we observed that incoherence can arise only if there are multiple actors that can read and write to caches and memory. Today, the most obvious collection of actors are the multiple cores on a single chip, but the cache coherence problem first arose in systems with a single core and direct memory access (DMA). A DMA controller is an actor that reads and writes memory under explicit system software control, typically at the page granularity. A DMA operation that reads memory should find the most recent version of each block, even if the block resides in a cache in state M or O. Similarly, a DMA operation that writes memory needs to invalidate all stale copies of the block.

It is straightforward to provide coherent DMA by adding a coherent cache to the DMA controller, and thus having DMA participate in the coherence protocol. In such a model, a

DMA controller is indistinguishable from a dedicated core, guaranteeing that DMA reads will always find the most recent version of a block and DMA writes will invalidate all stale copies.

However, adding a coherent cache to a DMA controller is undesirable for several reasons. First, DMA controllers have very different locality patterns than conventional cores, and they stream through memory with little, if any, temporal reuse. Thus, DMA controllers have little use for a cache larger than a single block. Second, when a DMA controller writes a block, it generally writes the entire block. Thus, fetching a block with a GetM is wasteful, since the entire data will be overwritten. Many coherence protocols optimize this case using special coherence operations. We could imagine adding a new GetM-NoData request to the protocols in this primer, which seeks M permission but expects only an acknowledgment message rather than a Data message. Other protocols use a special PutNewData message, which updates memory and invalidates all other copies including those in M and O.

DMA can also be made to work without hardware cache coherence, by requiring the operating system to selectively flush caches. For example, before initiating a DMA to or from a page P, the operating system could force all caches to flush page P using a protocol similar to TLB Shootdown (or using some other page flushing hardware support). This approach is inefficient, and thus generally only seen in some embedded systems, because the operating system must conservatively flush a page even if none of its blocks are in any cache.

9.1.6 MULTI-LEVEL CACHES AND HIERARCHICAL COHERENCE PROTOCOLS

Our baseline system assumes a single multicore chip with two levels of cache: private level-one data (L1) caches for each core and a shared last-level memory-side cache that holds both data and instructions (LLC). But many other combinations of chips and caches are possible. For example, Intel Nehalem and AMD Opteron processors support systems with multiple multicore chips as well as an additional level of private (per core) L2 caches. Figure 9.2 illustrates a system with two multicore processors, each having two cores with private L2 caches between the private L1s and shared LLC.

We next discuss multiple levels of caches on a single multicore chip, systems with multiple multicore processors, and hierarchical coherence protocols.

Multi-Level Caches

With multiple levels of caches, the coherence protocol must be sure to keep all of these caches coherent. Perhaps the most straightforward solution is to treat each cache completely independently. For example, the L1, L2, and LLC could each independently process all incoming coherence requests; this is the approach taken by the AMD Opteron [1].

However, we can also design the cache hierarchy such that not every cache needs to snoop every coherence request. As discussed in Section 8.6, a key design option is whether and which caches to make *inclusive*. An L2 is inclusive if it contains a superset of the blocks in the L1

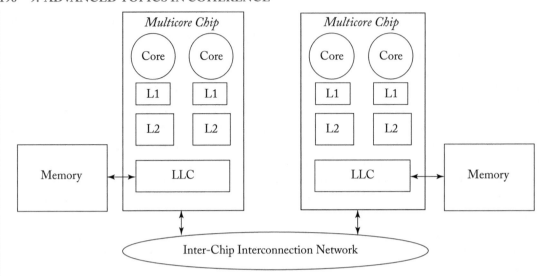

Figure 9.2: System with multiple multicore chips.

caches. Consider the case of an inclusive L2 when the L2 snoops a GetM for block B from another core. If B is not in the L2, then there is no need to also snoop the L1 caches because B cannot be in any of them. Thus, an inclusive L2 cache can serve as a filter that reduces the amount of coherence request traffic that must be snooped by the L1 caches. If instead B is in the L2, then B might also be in the L1 caches and then the L1 caches must also snoop the request. This is the approach taken by the AMD Bulldozer.

Inclusion's benefit—the reduction in L1 snoop bandwidth—must be traded off against the space wasted by redundant storage of inclusive blocks. The cache hierarchy can hold a greater number of distinct blocks if it is *exclusive* (i.e., if a block is in the L2 then it is not in the L1 caches) or non-inclusive (neither inclusive nor exclusive). Another reason not to provide inclusion is to avoid the complexity of maintaining inclusion (i.e., invalidating a block from the L1 when an L2 evicts that block).

Multiple Multicore Processors

Larger systems can be built by composing multiple multicore processor chips. While a full treatment of scalable systems is beyond the scope of this primer, we examine one key issue: how to use the LLC. In single-chip systems, the LLC is a memory-side cache logically associated with memory and thus can be largely ignored as far as coherence is concerned. In multi-chip systems, the LLC can alternatively be viewed as another level of the memory hierarchy. We present the options from the perspective of a given chip (the "local" chip) and its local memory; other chips are "remote" chips. The LLC could be used as:

- A memory-side cache that holds blocks recently requested from the local memory. The requests could be either from only the local chip or from both local and remote chips.

- A core-side cache that holds blocks recently requested by the cores on the chip. The blocks in the LLC could have homes on either this chip or other chips. In this case, the coherence protocol must usually operate among the LLCs and memories of the multiple multicore chips.

The LLC could also be used for both purposes, in a hybrid scheme. In a hybrid approach, the architects would have to decide how to allocate the LLC to these different demands.

Hierarchical Coherence Protocols

The protocols described in previous chapters are *flat* protocols, in that there was a single coherence protocol that every cache controller treats identically. However, once we introduce multiple levels of caches, we introduce the possible need for *hierarchical* coherence protocols.

Some systems are naturally hierarchical, including systems comprised of multiple multicore chips. Within each chip, there could be an intra-chip protocol, and there could be an inter-chip protocol across the chips. Coherence requests that can be satisfied by the intra-chip protocol do not interact with the inter-chip protocol; only when a request cannot be satisfied by another node on the chip does the request get promoted to the inter-chip protocol.

The choice of protocol at one level is largely independent of the choice at another level. For example, an intra-chip snooping protocol can be made compatible with an inter-chip directory protocol. Each chip would require a single directory controller that considers the entire chip to be a single node in the directory protocol. The inter-chip directory protocol could be identical to one of the directory protocols presented in Chapter 8, with the directory state naturally represented in a coarse fashion. Another possible hierarchical system could have directory protocols for both the intra- and inter-chip protocols, and the two directory protocols could be the same or even different from each other.

An advantage of hierarchical protocols for hierarchical systems is that it enables the design of a simple, potentially non-scalable intra-chip design for the commodity chip. When designing a chip, it would be beneficial to not have to design a single protocol that scales to the largest possible number of cores that could exist in a system. Such a protocol is likely to be overkill for the vast majority of systems that are comprised of a single chip.

There are numerous examples of hierarchical protocols for hierarchical systems. The Sun Wildfire prototype [5] connects multiple snooping systems together with a higher level directory protocol. The AlphaServer GS320 [4] has two levels of directory protocols, one within each quad-processor block and another across these blocks. The Stanford DASH machine [10] consisted of multiple snooping systems connected by a higher level directory protocol.

In a system with hundreds or thousands of cores, it might not make much sense to have a single coherence protocol. The system may be more likely to be divided, either statically or dynamically, into domains that each run a separate workload or separate virtual machine. In such a system, it may make sense to implement a hierarchical protocol that optimizes for intra-domain sharing yet still permits inter-domain sharing. Recent work by Marty and Hill [13] superposes a hierarchical coherence protocol on top of a multicore chip with a flat design. This design enables the common case—intra-domain sharing—to be fast while still allowing sharing across domains.

9.2   PERFORMANCE OPTIMIZATIONS

There is a long history of research into optimizing the performance of coherence protocols. Rather than present a high-level survey, we focus on two optimizations that are largely independent of whether the underlying coherence protocol is snooping or directory. Why these two optimizations? Because they can be effective and they illustrate the kinds of optimizations that are possible.

9.2.1   MIGRATORY SHARING OPTIMIZATION

In many multithreaded programs, a common phenomenon is *migratory sharing*. For example, one core may read and then write a data block, then a second core may read and write it, and so on. This pattern commonly arises from critical sections (e.g., the lock variable itself) in which the data block migrates from one core to another. In a typical protocol, each core performs a GetS transaction to read the data and then a subsequent GetM transaction to get write permission for the same block. However, if the system can predict that the data conforms to a migratory sharing pattern, cores could get an exclusive copy of the block when they first read it, thus reducing both the latency and bandwidth to access the data [2, 15, 17]. The migratory optimization is similar to the E state optimization, except that it also needs to return an exclusive copy when the block is in state M in some cache, not just when the block is in state I in all caches.

There are two basic approaches to optimizing migratory sharing. First, one can use some hardware predictor to predict that a particular block exhibits a migratory sharing pattern and issue a GetM rather than a GetS on a load miss. This approach requires no change to the coherence protocol but introduces a few challenges:

- Predicting migratory sharing: we must design a hardware mechanism to predict when a block is undergoing migratory sharing. A simple approach is to use a table to record which blocks were first obtained with a GetS and then subsequently written, requiring a GetM. On each load miss, the coherence controller could consult the predictor to determine whether a block exhibits a migratory sharing pattern. If so, it could issue a GetM request, rather than a GetS.

- Mispredictions: if a block is not migrating, then this optimization can hurt performance. Consider the extreme case of a system in which cores issue only GetM requests. Such a system would never permit multiple cores to share a block in a read-only state.

Alternatively, we can extend the coherence protocol with an additional Migratory M (MM) state. The MM state is equivalent to the state M, from a coherence perspective (i.e., dirty, exclusive, owned), but it indicates that the block was obtained by a GetS in response to a predicted migratory sharing pattern. If the local core proceeds to modify a block in MM, re-inforcing the migratory sharing pattern, it changes the block to state M. If a core in state M receives a GetS from another core, it predicts that the access will be migratory and sends exclusive data (invalidating its own copy). If the Other-GetS finds the block in state MM, the migratory pattern has been broken and the core sends a shared copy and reverts to S (or possibly O). Thus, if many cores make GetS requests (without subsequent stores and GetM requests), all cores will receive S copies.

Migratory sharing is just one example of a phenomenon that, if detected, can be exploited to improve the performance of coherence protocols. There have been many schemes that target specific phenomena, as well as more general approaches to predicting coherence events [14].

9.2.2 FALSE SHARING OPTIMIZATIONS

One performance problem that can plague coherence protocols is *false sharing*. False sharing occurs when two cores are reading and writing different data that happen to reside on the same cache block. Even though the cores are not actually sharing the data on the block (i.e., the sharing is false), there can be a significant amount of coherence traffic between the cores for the block. This coherence traffic hurts performance when a core is waiting for coherence permissions to access a block, and it increases the load on the interconnection network. The likelihood of false sharing occurring is a function of the block size—a larger block can hold more unrelated pieces of data and thus larger blocks are more prone to false sharing—and the workload. There are at least two optimizations to mitigate the impact of false sharing.

Sub-Block Coherence
Without reducing the block size, we can perform coherence at a finer, sub-block granularity [7]. Thus, it is possible for a block in a cache to have different coherence states for different sub-blocks. Sub-blocking reduces false sharing, but it requires extra state bits for each block to hold the sub-block states.

Speculation
An architect can develop a hardware mechanism to predict when a block that is invalid in a cache is the victim of false sharing [6]. If the predictor believes the block is invalid due to false sharing, the core can speculatively use the data in the block until it obtains coherence permissions to the block. If the prediction was correct, this speculation overcomes the latency penalty of false sharing, but it does not reduce the traffic on the interconnection network.

9.3   MAINTAINING LIVENESS

In Chapter 2, we defined coherence and the invariants that must be maintained by a coherence protocol. These invariants are *safety* invariants; if these invariants are maintained, then the protocol will never allow unsafe (incorrect) behavior. Facetiously, it is easy to provide safety because an unplugged computer never does anything incorrect! The key is to provide both safety and *liveness*, where providing liveness requires the prevention of three classes of situations: deadlock, livelock, and starvation.

9.3.1   DEADLOCK

As discussed briefly in Section 8.2.3, deadlock is the situation in which two or more actors wait for each other to perform some action, and thus never make progress. Typically, deadlock results from a cycle of resource dependences. Consider the simple case of two nodes A and B and two resources X and Y. Assume A holds resource Y and B holds resource X. If A requests X and B requests Y, then unless one node relinquishes the resource it already holds, these two nodes will deadlock. We illustrate this cyclical dependence graph in Figure 9.3. More generally, deadlock can result from cycles involving many nodes and resources. Note that partial deadlocks (e.g., the simple case of deadlock between only nodes A and B) can quickly become complete system deadlocks when other nodes wait for deadlocked nodes to perform some action (e.g., node C requests resource X).

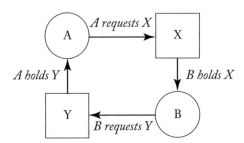

Figure 9.3: Example of deadlock due to cyclical resource dependences. Circles are nodes and squares are resources. An arc that ends at a resource denotes a request for that resource. An arc that starts at a resource denotes the holder of that resource.

Protocol Deadlocks

In coherence protocols, deadlock can arise at the protocol level, at cache resource allocation, and in the network. Protocol deadlocks arise when a coherence controller waits for a message that will never be sent. For example, consider a (buggy) directory protocol that does not wait for a Put-Ack after sending a PutS, and instead immediately transitions to state I. If the directory controller sends an Inv (e.g., in response to core C1's GetM request) to core C0 at the same time that core C0 sends a PutS to the directory, then C1 will never get an Inv-Ack from core C0 and will deadlock waiting for it. Such deadlocks represent protocol errors and usually arise from untested race conditions.

Cache Resource Deadlocks

Cache resource deadlocks arise when a cache controller must allocate a resource before performing some action. These deadlocks typically arise either when handling another core's request or on writebacks. For example, consider a cache controller that has a set of shared buffers (e.g., transaction buffer entries, or TBEs) that may be allocated *both* when the core initiates a coherence request and when servicing another core's request. If the core issues enough coherence requests to allocate all the buffers, then it cannot process another core's request until it completes one of its own. If all cores reach this state, then the system deadlocks.

Protocol-Dependent Network Deadlocks

There are two causes of network deadlocks: deadlocks due to buggy routing algorithms, which are independent of the types of messages and the coherence protocol, and network deadlocks that arise because of the particular messages being exchanged during coherence protocol operation. We focus here on this latter category of protocol-dependent network deadlocks. Consider a directory protocol in which a request message may lead to a forwarded request and a forwarded request may lead to a response. The protocol must ensure three invariants to avoid cyclic dependences and thus deadlock.

Sidebar: Virtual Networks

Instead of using physically distinct networks, we can use distinct *virtual networks*. Consider two cores that are connected with a single point-to-point link. At the end of each link is a FIFO queue to hold incoming messages before the receiving core can process them. This single network is shown in Fig. 9.4a. To add another physical network, as shown in Fig. 9.4b, we duplicate the links and the FIFO queues. Requests travel on one physical network, and replies travel on the other physical network.

To avoid the cost of replicating the links and switches (switches not shown in figures), we can add a virtual network, as illustrated in Fig. 9.4c. The only cost of a virtual network is an additional FIFO queue at each switch and endpoint in the network. Adding the second virtual network in this example allows requests to not get stuck behind replies.

Virtual networks are related to virtual channels [3], and some papers use the terms interchangeably. However, we prefer to distinguish between them because they address different types of deadlocks. Virtual networks prevent messages of different classes from blocking each other and thus avoid message-level deadlocks.

Virtual channels are used at the network level to avoid deadlocks due to routing, regardless of the message types. To avoid routing deadlock, messages travel on multiple virtual channels (e.g., a message traveling west in a 2D torus might be required to use virtual channel 2). A virtual channel, like a virtual network, is implemented as an extra FIFO queue at each switch and end point in the network. Virtual channels are orthogonal to virtual networks; each virtual network may have some number of virtual channels to avoid routing deadlock.

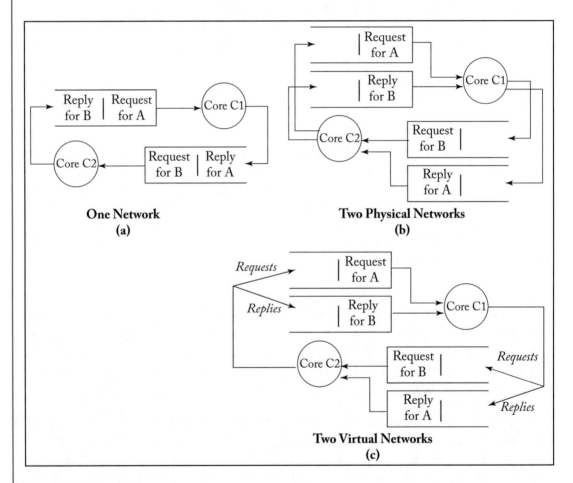

Figure 9.4: Virtual networks.

- As explained in Section 8.2.3, each message class must travel on its own network. The networks may be physical or virtual (see sidebar on "virtual networks"), but the key is avoiding situations in which a message of one class becomes stuck behind a message of another class in a FIFO buffer. In this example, requests, forwarded requests, and responses all travel on separate networks. A coherence controller thus has three incoming FIFOs, one for each network.

- Message classes must have a dependence order. If a message of class A can cause a coherence controller to issue a message of class B, then a coherence controller may not stall the processing of an incoming class B message while waiting for a class A message. In the directory example, a coherence controller cannot stall the processing of a forwarded request while waiting for a request to arrive, nor can it stall the processing of a response while waiting for a forwarded request.

- The last message class in this chain of dependences—the response message, in this directory example—must always be "sunk." That is, if a coherence controller receives a response message, there must be no message class than can prevent it from being removed from its incoming queue.

These three invariants eliminate the possibility of a cycle. Even though a request can be stalled while waiting for responses or forwarded requests, every request will eventually be processed because the number of responses and forwarded requests is bounded by the number of outstanding transactions.

9.3.2   LIVELOCK

Livelock is a situation in which two or more actors perform actions and change states, yet never make progress. Livelock is a special case of starvation, discussed next. Livelock occurs most frequently in coherence protocols that use negative acknowledgment messages (NACKs). A node may issue a coherence request, but receive a NACK, prompting a retry. If contention or some repeatable race with another node causes this case to recur indefinitely, then the nodes livelock. The protocols in this primer do not use NACKs, so we focus on another well-known livelock scenario involving coherence permissions that can arise in these protocols.

This cause of livelock is the so-called "window of vulnerability" problem [8], an example of which is illustrated in Table 9.1. Consider a snooping protocol in which Core C1 issues a GetS request for block B and changes B's state to IS^{AD} (in I, going to S, waiting for Own-GetS and data). At some point later, C1 observes its Own-GetS on the bus and changes B's state to state IS^{D}. Between when B goes to state IS^{D} and when C1 receives the data response, C1 is vulnerable to observing a GetM request for B from another core on the bus. In an optimized protocol, like the protocol in Section 7.5.5, if a GetM arrives for B in state IS^{D}, C1 will change B's state to $IS^{D}I$. In this transient state, when C1 later receives the data response, C1 changes B's state to I. Because C1 cannot perform a load to a block in I, it must issue another GetS for B. However, this next GetS is just as susceptible to the window of vulnerability problem, and thus C1 may never make forward progress. The core is still active, and thus the system is not deadlocked, but it never makes forward progress. Somewhat perversely, this situation is most

Table 9.1: Livelock example for Core C1 trying to load block B in a snooping protocol

Cycle	Event (all for block B)	Core C1's State for B
0	*Initial state*	I
1	Load request, issue GetS to bus	IS^{AD}
2	Observe Own-GetS on bus	IS^{D}
3	Observe Other-GetM on bus	$IS^{D}I$
4	Receive data for Own-GetS	I
5	Re-issue GetS to bus	IS^{AD}
6	Observe Own-GetS on bus	IS^{D}
7	Observe Other-GetM on bus	$IS^{D}I$
8	Receive data for Own-GetS	I
9	Etc. (never completing the load)	

likely to arise for highly contended blocks, which means that most or all of the cores are likely to simultaneously be stuck and thus the system can livelock.

This window of vulnerability can be closed by requiring that C1 perform at least one load when it receives the data response. This load *logically appears to occur* at the time at which C1's GetS is ordered (e.g., on the bus in a snooping protocol) and thus does not violate coherence. However, if certain conditions are not satisfied, performing this load could violate the memory consistency model. Satisfying these conditions is sometimes known as the Peekaboo problem, and we discuss it in more detail in the sidebar.

Sidebar: Peekaboo Coherence Problem

Table 9.2 illustrates what is sometimes called the Peekaboo Coherence problem. In this example, the locations A and B are initially zero, core C0 writes A first and then B, and core C1 reads B first and then A. Under both the SC and TSO memory consistency models, the only illegal outcome is r1=1 and r2=0. This example execution uses the optimized directory protocol from Section 8.7.2, but elides the directory controller's actions (which are not pertinent to the example). PrefetchS is the one new operation in this example, which issues a GetS request if a readable block does not already reside in the cache.

The Peekaboo problem arises when a block is prefetched, invalidated before permission is received, and then a demand reference (demanded by a load or store, not a prefetch) occurs. If we perform the demand reference when the prefetched but already invalidated Data arrives, then the demand reference is effectively ordered at the time the block was invalidated. In this example, C1's load A is effectively ordered at time 4 (when C1 receives the

Inv for block A), while C1's earlier (in program order) load B is ordered at time 7. Reordering these two loads violates both SC and TSO. Note that this problem can arise whether the prefetch operation results from an explicit prefetch instruction, hardware prefetcher, or speculative execution. This problem can also arise in optimized snooping protocols, such as the one in Section 7.5.5.

The simplest solution to the Peekaboo problem is to perform the load in the window of vulnerability *if and only if* that load was the oldest load in program order when the coherence request was first issued. In other words, the Peekaboo load is allowed to access the invalidated data only if all previous loads and stores before the Peekaboo load in program order were already performed before the coherence request for the Peekaboo load was first issued [11]. A complete analysis of why this solution is sufficient can be found in Manerkar et al. [11], but intuitively the problem cannot arise if a core issues coherence requests one at a time in the order of demand misses.

Table 9.2: Example of Peekaboo Coherence problem

Time	Core C0 A = B = 0 initially Store A = 1 Store B = 1	Core C1 Prefetch A (prefetch for read-only access) Load r1 = B Load r2 = A
0	A:M[0] B:M[0]	A:I B:I
1		A: prefetchS miss, issue GetS/IS^D
2	A: Receive Fwd-GetS, send Data[0]/S	
3	A: store miss; issue GetM/SM^{AD}	
4	A: receive Data[0](ack=1)/SM^A	A: receive Inv, send Inv-Ack/IS^{D}I
5	A: receive Inv-Ack, perform store/M[1]	
6	B: store hit/M[1]	
7		B: load miss, issue GetS/IS^D
8	B: receive Fwd-GetS, send Data[1]/S[1]	
9		B: receive Data[1], perform load r1=1/S[1]
10		A: load miss, stall/IS^{D}I
11		**A: receive Data[0], perform load r2=0/I**
	Core C1 observes A = 0 and B = 1, effectively reordering the loads.	

This same window of vulnerability exists for stores to blocks in IM^DS, IM^DSI, or IM^DI. In these cases, the store to the block is never performed because the block's state at the end of the transaction is either I or S, which is insufficient for performing a store. Fortunately, the same solution we presented for the load in IS^DI applies to stores in these states. A core that issues a GetM must perform at least one store when it receives the data, and the core must forward this newly written data to the other core(s) that requested the block in S and/or M between when it observes its own GetM and when it receives the data in response to its GetM. Note that the same restriction needed to avoid the Peekaboo problem still applies: namely, perform the store if and only if the store was the oldest load or store in program order at the time the coherence request was issued.

9.3.3   STARVATION

Starvation is a situation in which one or more cores fail to make forward progress while other cores are still actively making forward progress. The cores not making progress are considered to be starved. There are several root causes of starvation, but they tend to fall into two categories: unfair arbitration and incorrect use of negative acknowledgments.

Starvation can arise when at least one core cannot obtain a critical resource because the resource is always obtained or held by other cores. A classic example of this is an unfair bus arbitration mechanism in a bus-based snooping protocol. Consider a bus in which access to the bus is granted in a fixed priority order. If Core C1 wishes to make a request, it can make a request. If C2 wishes to make a request, it may make the request only if C1 has not first requested the bus. C3 must defer to C1 and C2, etc. In such a system, a core with a low priority may never gain permission to make a request and will thus starve. This well-known problem also has a well-known solution: fair arbitration.

The other main class of starvation causes is the incorrect use of negative acknowledgments (NACKs) in coherence protocols. In some protocols, a coherence controller that receives a coherence request may send a NACK to the requestor (often used in verb form as "the controller NACKed the request"), informing the requestor that the request was not satisfied and must be re-issued. NACKs are generally used by protocols to simplify situations in which there is another transaction in progress for the requested block. For example, in some directory protocols, the directory can NACK a request if the requested block is already in the midst of a transaction. Solving these protocol race conditions with NACKs appears, at least at first blush, to be conceptually easier than designing the protocol to handle some of these rare and complicated situations. However, the challenge is ensuring that a NACKed request eventually succeeds. Guaranteeing a lack of starvation, regardless of how many cores are requesting the same block at the same time, is challenging; one of the authors of this primer confesses to having designed a protocol with NACKs that led to starvation.

9.4   TOKEN COHERENCE

Until fairly recently, coherence protocols could be classified as either snooping or directory or perhaps a hybrid of the two. There were many variants of each class and several hybrids, but protocols were fundamentally some combination of snooping and directory. In 2003, Martin et al. proposed a third protocol classification called *Token Coherence* [12]. There are two key ideas behind Token Coherence (TC).

TC protocols associate tokens with each block instead of state bits. There is a fixed number of tokens per block, and the cores can exchange—but not create or destroy—these tokens. A core with one or more tokens for a block can read the block, and a core with all of the tokens for a block can read or write to the block.

A TC protocol consists of two distinct parts: a correctness substrate and a performance protocol. The correctness substrate is responsible for ensuring safety (tokens are conserved) and liveness (all requests are eventually satisfied). The performance protocol specifies what a cache controller does on a cache miss. For example, in the TokenB performance protocol, all coherence requests are broadcast. In the TokenM protocol, coherence requests are multicast to a predicted set of sharers.

Token Coherence subsumes snooping and directory protocols, in that snooping and directory protocols can be interpreted as TC protocols. For example, an MSI snooping protocol is equivalent to a TC protocol with a broadcast performance protocol. The MSI states are equivalent to a core having all/some/none of the tokens for a block.

9.5   THE FUTURE OF COHERENCE

Almost since coherence's invention, some have predicted that it will soon go away because it adds hardware cost to store extra state, send extra messages, and verify that all is correct. However, we predict that coherence will remain commonly implemented because, in our judgment the software cost of dealing with incoherence is often substantial and borne by a broader group of software engineers rather than the few hardware designers that confront implementing coherence. One piece of evidence for this claim is the adoption of coherence in recent heterogeneous systems, as we discuss in the next chapter. Differently from the protocols we discussed in Chapters 6–9, however, we will see that heterogeneous systems are suited to consistency-directed rather than consistency-agnostic coherence protocols.[2]

9.6   REFERENCES

[1] P. Conway and B. Hughes. The AMD Opteron northbridge architecture. *IEEE Micro*, 27(2):10–21, March/April 2007. DOI: 10.1109/mm.2007.43. 195

---

[2]Recall that in a consistency-directed coherence protocol (Section 2.3) there is no clean separation of coherence from consistency.

[2] A. L. Cox and R. J. Fowler. Adaptive cache coherency for detecting migratory shared data. In *Proc. of the 20th Annual International Symposium on Computer Architecture*, pp. 98–108, May 1993. DOI: 10.1109/isca.1993.698549. 198

[3] W. J. Dally. Virtual channel flow control. *IEEE Transactions on Parallel and Distributed Systems*, 3(2):194–205, March 1992. DOI: 10.1109/71.127260. 201

[4] K. Gharachorloo, M. Sharma, S. Steely, and S. V. Doren. Architecture and design of AlphaServer GS320. In *Proc. of the 9th International Conference on Architectural Support for Programming Languages and Operating Systems*, pp. 13–24, November 2000. DOI: 10.1145/378993.378997. 197

[5] E. Hagersten and M. Koster. WildFire: A scalable path for SMPs. In *Proc. of the 5th IEEE Symposium on High-Performance Computer Architecture*, pp. 172–81, January 1999. DOI: 10.1109/hpca.1999.744361. 197

[6] J. Huh, J. Chang, D. Burger, and G. S. Sohi. Coherence decoupling: Making use of incoherence. In *Proc. of the 11th International Conference on Architectural Support for Programming Languages and Operating Systems*, October 2004. DOI: 10.1145/1024393.1024406. 199

[7] M. Kadiyala and L. N. Bhuyan. A dynamic cache sub-block design to reduce false sharing. In *Proc. of the International Conference on Computer Design*, 1995. DOI: 10.1109/iccd.1995.528827. 199

[8] J. Kubiatowicz, D. Chaiken, and A. Agarwal. Closing the window of vulnerability in multiphase memory transactions. In *Proc. of the 5th International Conference on Architectural Support for Programming Languages and Operating Systems*, pp. 274–84, October 1992. DOI: 10.1145/143365.143540. 203

[9] H. Q. Le et al. IBM POWER6 microarchitecture. *IBM Journal of Research and Development*, 51(6), 2007. DOI: 10.1147/rd.516.0639.

[10] D. Lenoski, J. Laudon, K. Gharachorloo, A. Gupta, and J. Hennessy. The directory-based cache coherence protocol for the DASH multiprocessor. In *Proc. of the 17th Annual International Symposium on Computer Architecture*, pp. 148–59, May 1990. DOI: 10.1109/isca.1990.134520. 197

[11] Y. A. Manerkar, D. Lustig, M. Pellauer, and M. Martonosi. CCICheck: using μhb graphs to verify the coherence-consistency interface. In *Proc. of the 48th International Symposium on Microarchitecture*, 2015. DOI: 10.1145/2830772.2830782. 205

[12] M. M. K. Martin, M. D. Hill, and D. A. Wood. Token coherence: Decoupling performance and correctness. In *Proc. of the 30th Annual International Symposium on Computer Architecture*, June 2003. DOI: 10.1109/isca.2003.1206999. 191, 207

[13] M. R. Marty and M. D. Hill. Virtual hierarchies to support server consolidation. In *Proc. of the 34th Annual International Symposium on Computer Architecture*, June 2007. DOI: 10.1145/1250662.1250670. 198

[14] S. S. Mukherjee and M. D. Hill. Using prediction to accelerate coherence protocols. In *Proc. of the 25th Annual International Symposium on Computer Architecture*, pp. 179–90, June 1998. DOI: 10.1109/isca.1998.694773. 199

[15] J. Nilsson and F. Dahlgren. Improving performance of load-store sequences for transaction processing workloads on multiprocessors. In *Proc. of the International Conference on Parallel Processing*, pp. 246–55, September 1999. DOI: 10.1109/icpp.1999.797410. 198

[16] B. F. Romanescu, A. R. Lebeck, D. J. Sorin, and A. Bracy. UNified instruc-tion/translation/data (UNITD) coherence: One protocol to rule them all. In *Proc. of the 15th International Symposium on High-Performance Computer Architecture*, January 2010. DOI: 10.1109/hpca.2010.5416643. 192

[17] P. Stenström, M. Brorsson, and L. Sandberg. An adaptive cache coherence protocol opti-mized for migratory sharing. In *Proc. of the 20th Annual International Symposium on Com-puter Architecture*, pp. 109–18, May 1993. DOI: 10.1109/isca.1993.698550. 198

[18] P. J. Teller. Translation-lookaside buffer consistency. *IEEE Computer*, 23(6), pp. 26–36, June 1990. DOI: 10.1109/2.55498. 192

CHAPTER 10

Consistency and Coherence for Heterogeneous Systems

This is the age of specialization. Today's server, mobile, and desktop processors contain not only conventional CPUs, but also various flavors of accelerators. The most prominent among the accelerators are Graphics Processing Units (GPUs). Other types of accelerators including Digital Signal Processors (DSPs), AI accelerators (e.g., Apple's Neural Engine, Google's TPU), cryptographic accelerators, and field-programmable-gate-arrays (FPGAs) are also becoming common.

Such heterogeneous architectures, however, pose a programmability challenge. How to synchronize and communicate within and across the accelerators? And, how to do this efficiently? One promising trend is to expose a global shared memory interface across the CPUs and the accelerators. Recall that shared memory helps not only with programmability by providing an intuitive load-store interface for communication, but also helps in reaping the benefits of locality via programmer-transparent caching.

CPUs, GPUs, and other accelerators can either be tightly integrated and actually share the same physical memory, as is the case in mobile systems-on-chips, or they may have physically independent memories with a runtime that provides a logical abstraction of shared memory. Unless specified otherwise, we assume the former. As shown in Figure 10.1, each of the CPUs, GPUs, and accelerators may have multiple cores with private per-core L1s and a shared L2. The CPUs and the accelerators also may share a memory-side last-level cache (LLC) that, unless specified otherwise, is non-inclusive. (Recall that a memory-side LLC does not pose coherency issues). The LLC also serves as an on-chip memory controller.

Shared memory automatically raises questions. What is the consistency model? How is the consistency model (efficiently) enforced? In particular, how are the caches within an accelerator (the L1s) and across the CPUs and accelerators (L1s and L2s) kept coherent?

In this chapter, we discuss potential answers to these questions in this fast moving area. We first study consistency and coherence within an accelerator, focusing on GPUs. We then consider consistency and coherence across the accelerators and the CPU.

10.1 GPU CONSISTENCY AND COHERENCE

We start this section by briefly summarizing early GPU architectures and their programming model. This will help us appreciate the design choices with regard to consistency and coherence in such GPUs, which were primarily targeted toward graphics workloads. We then discuss the

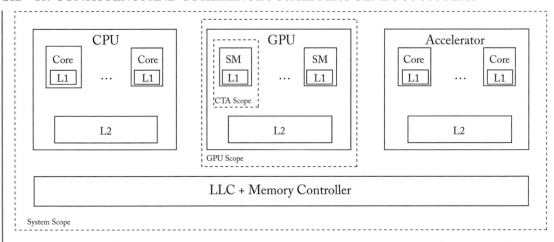

Figure 10.1: System model of a heterogeneous system-on-chip containing a CPU, GPU, and accelerator. SM refers to a streaming multi-processor; CTA refers to cooperative thread array.

trend of using GPU-like architectures to run general-purpose workloads in so-called General-Purpose Graphics Processing Units (GPGPUs) [5]. Specifically, we discuss the new demands that GPGPUs place on consistency and coherence. We then discuss in detail some of the recent proposals for fulfilling these demands.

10.1.1 EARLY GPUS: ARCHITECTURE AND PROGRAMMING MODEL

Early GPUs were primarily tailored toward embarrassingly parallel graphics workloads. Roughly speaking, the workloads involve computing independently each of the pixels that constitute the display. Thus, the workloads are characterized by a very high degree of data-parallelism and low degrees of data sharing, synchronization, and communication.

GPU Architecture

To exploit the abundance of parallelism, GPUs typically have tens of cores called Streaming Multiprocessors (SMs)[1], as shown in Figure 10.1. Each SM is highly multithreaded, capable of running on the order of a thousand threads. The threads mapped to an SM share the L1 cache and a local scratchpad memory (not shown). All of the SMs share an L2 cache.

In order to amortize the cost of fetching and decoding instructions for all of these threads, GPUs typically execute threads in groups called warps.[2] All of the threads in a warp share the program counter (PC) and the stack, but can still execute independent thread-specific paths using mask-bits that specify which threads among the warps are active and which threads should not execute. This style of parallelism is known as "Single-Instruction-Multiple-Threads"

---

[1]SM in NVIDIA parlance. Also known as Compute Units (CUs) in AMD parlance.
[2]Warp in NVIDIA parlance. Also known as Wavefront in AMD parlance.

(SIMT). Because all of the threads in a warp share the PC, traditionally the threads in a warp were scheduled together. Recently, however, GPUs are starting to allow for threads within a warp to have independent PCs and stacks, and consequently allow for the threads to be independently scheduled. For the rest of the discussion, we assume that threads can be independently scheduled.

Because graphics workloads do not share data or synchronize frequently—synchronization and communication typically happening infrequently at a coarser level of granularity—early GPUs chose not to implement hardware cache coherence for the L1 caches. (They did not enforce the SWMR invariant.) The absence of hardware cache coherence makes synchronization a trickier prospect for the programmer, as we see next.

GPU Programming model

Analogously to the CPU instruction set architecture (ISA), the GPUs also have virtual ISAs. NVIDIA's virtual ISA, for example, is known as Parallel Thread Execution (PTX). In addition, there are higher-level language frameworks. Two such frameworks are especially popular: CUDA and OpenCL. The high-level programs in these frameworks are compiled down to virtual ISAs, which in turn are translated into native binaries at kernel installation time. A kernel refers to a unit of work offloaded by the CPU onto the GPU and typically consists of a large number of software threads.

GPU virtual ISAs as well as language frameworks choose to expose the hierarchical nature of GPU architectures to the programmer via a thread hierarchy known as *scopes* [15]. In contrast to CPU threads where all threads are "equal", GPU threads from a kernel are grouped into clusters called Cooperative Thread Arrays (CTA).[3] A *CTA* scope refers to the set of threads from the same CTA. Such threads are guaranteed to be mapped to the same SM and hence share the same L1. Thus, the CTA scope also implicitly refers to the memory hierarchy level (L1[4]) that all of these threads share. A *GPU* scope refers to the set of threads from the same GPU. These threads could be from the same or different CTAs from that GPU. All of these threads share the L2. Thus, the GPU scope implicitly refers to the L2. Finally, *System* scope refers to the set of all threads from the whole system. These threads can be from the CPU, GPU, or other accelerators that constitute the system. All of these threads may share a cache (LLC). Thus, the System scope implicitly refers to the LLC, or unified shared memory if there is no shared LLC.

Why expose the thread and memory hierarchy to the software? In the absence of hardware cache coherence, this allows for programmers and hardware to cooperatively achieve synchronization efficiently. Specifically, if the programmer ensures that two synchronizing threads are within the same CTA, the threads can synchronize and communicate efficiently via the L1 cache.

---

[3]CTA in PTX parlance. Also known as Workgroups in OpenCL and as thread blocks in CUDA.
[4]and the local scratchpad memory, if present

How can two threads belonging to different CTAs from the same GPU kernel synchronize? Somewhat surprisingly, early GPU consistency models did not explicitly allow this. In practice, though, programmers could achieve GPU-scoped synchronization via specially crafted loads and stores that bypass the L1 and synchronize at the shared L2. Needless to say, this was causing a great deal of confusion among the programmers [8]. In the next section, we explore GPU consistency in detail with examples.

GPU Consistency

GPUs support relaxed memory consistency models. Like relaxed consistency CPUs, GPUs enforce only the memory orderings indicated by the programmer, e.g., via FENCE instructions.

However, because of the absence of hardware cache coherence in GPUs, the semantics of a FENCE instruction is different. Recall that a FENCE instruction in a multicore CPU ensures that loads and stores before the FENCE are performed, with respect to all threads, before the loads and stores following the FENCE. A GPU FENCE provides similar semantics, but only with respect to other threads belonging to the same CTA. A corollary of this is that GPU stores lack atomicity. Recall that store atomicity (Section 5.5.2) mandates that a thread's store is logically seen by all other threads at once. In GPUs, however, a store may become visible to a thread belonging to the same CTA before other threads.

Consider the message-passing example shown in Table 10.1 where the programmer intends for load Ld2 to read NEW and not the old value of 0. How does one ensure this?

Table 10.1: Message passing example. T1 and T2 belong to the same CTA.

Thread T1	Thread T2	Comments
St1: St data = NEW;	Ld1: Ld r1 = flag;	// Initially data and flag are 0.
FENCE;	B1: if (r1 ≠ SET) goto Ld1;	
St2: St flag = SET;	FENCE;	
	Ld2: Ld r2 = data;	Can r2==0?

First of all, note that without the two FENCE instructions, Ld2 can read the old value of 0—GPUs, being relaxed, may perform loads and stores out of order.

Now, let us suppose that the two threads T1 and T2 belong to the same CTA, and are hence mapped to the same SM. At the microarchitectural level, a GPU FENCE works similarly to a FENCE in XC, as we discussed in Chapter 5. The reorder unit ensures that the Load/Store→FENCE and FENCE→Load/Store orderings are enforced. Because T1 and T2 share an L1, honoring the above rules is sufficient to ensure that the two stores from T1 are made visible to T2 in program order, ensuring that load L2 reads NEW.

Table 10.2: Message passing example. T1 and T2 belong to different CTAs.

Thread T1	Thread T2	Comments
St1: St.GPU data = NEW; FENCE; St2: St.GPU flag = SET;	Ld1: Ld.GPU r1 = flag; B1: if (r1 ≠ SET) goto Ld1; FENCE; Ld2: Ld.GPU r2 = data;	// Initially data and flag are 0. Can r2==0?

On the other hand, if the two threads T1 and T2 belong to different CTAs, and hence are mapped to different SMs (SM1 and SM2), it is possible for load Ld2 to read 0. To see how, consider the following sequence of events.

- Initially, both data and flag are cached in both L1s.

- St1 and St2 perform in program order, writing NEW and SET, respectively, in SM1's L1 cache.

- The cache line holding flag is evicted from SM1's L1, and flag=SET is written to the L2.

- The cache line holding flag in SM2's L1 is evicted.

- The load Ld1 performs and misses in SM2's L1, so the line from the L2 is fetched and reads SET.

- The load Ld2 performs, hits in SM2's L1, and reads 0.

Thus, although the FENCE ensures that the two stores are written to SM1's L1 in the correct order, they may become visible to T2 in a different order in the absence of hardware cache coherence. That is why early GPU programming manuals explicitly disallow this type of inter-CTA synchronization between threads of the same kernel.

Is inter-CTA synchronization then impossible to achieve? In practice, a workaround is possible, leveraging special load and store instructions that directly target specific levels of the memory hierarchy. As we can see in Table 10.2, the two stores in T1 explicitly write to the GPU scope (i.e., to the L2) bypassing the L1. Likewise, the two loads from T2 bypass the L1 and read directly from the L2. Thus, the two threads of GPU scope, T1 and T2, are explicitly made to synchronize at the L2 by using loads and stores that bypass the L1.

However, there are problems with the above workaround—the primary issue being performance inefficiency owing to loads and stores bypassing the L1. In this simple example, the two variables in question must be communicated across SMs, so bypassing is a necessary evil. However, bypassing may be problematic with more variables, only some of which are communicated across SMs. In such a situation, the programmer is confronted with the onerous task of carefully directing loads/stores to the appropriate memory hierarchy level in order to make effective use of the L1.

Flashback to Quiz Question 8: GPUs do not support hardware cache coherence. Therefore, they are unable to enforce a memory consistency model. *True or false?*
Answer: *False!* Early GPUs did not support cache coherence in hardware, yet supported scoped relaxed consistency models.

Summary: Limitations and Requirements

Early GPUs were primarily targeted toward embarrassingly parallel workloads that neither synchronized nor shared data frequently. Therefore, such GPUs chose not to support hardware cache coherence for keeping the local caches coherent, at the cost of a scoped memory consistency model that permitted only intra-CTA synchronization. More flexible inter-CTA synchronization was either too inefficient or placed a huge burden on programmers.

Programmers are starting to use GPUs for general purpose workloads. Such workloads tend to involve relatively frequent fine-grained synchronization and more general sharing patterns. Thus, it is desirable for a GPGPU to have:

- a rigorous and intuitive memory consistency model that permits synchronization across all threads; and

- a coherence protocol that enforces the consistency model while allowing for efficient data sharing and synchronization, while at the same time keeping the simplicity of the conventional GPU architecture, as GPUs will still cater to graphic workloads predominantly.

10.1.2 BIG PICTURE: GPGPU CONSISTENCY AND COHERENCE

We already outlined the desired properties for GPGPU consistency and coherence. One straightforward approach to meeting these demands is to use a multicore CPU-like approach to coherence and consistency, i.e., use one of the consistency-agnostic coherence protocols (that we covered at length in Chapters 6–9) to ideally enforce a strong consistency model such as sequential consistency (SC). Despite ticking almost all boxes—the consistency model is certainly intuitive (without the notion of scopes) and the coherence protocol allows for the efficient sharing of data—the approach is ill-suited for a GPU. There are two major reasons for this [30].

First, a CPU-like coherence protocol that invalidates sharers upon a write would incur a high traffic overhead in the GPU context. This is because the aggregate capacity of the local caches (L1s) in GPUs is typically comparable to, or even greater than, the size of the L2. The NVIDIA Volta GPU, for example, has an aggregate capacity of about 10 MB of L1 cache and only 6 MB of L2 cache. A standalone inclusive directory would not only incur a large area overhead owing to the duplicate tags, but also significant complexity because the directory would need to be highly associative. On the other hand, an embedded inclusive directory would result in a significant amount of recall traffic upon L2 evictions, given the size of the aggregate L1s.

Second, because GPUs maintain thousands of active hardware threads, there is a need to track a correspondingly high number of coherence transactions, which would cost significant hardware overhead.

Without writer-initiated invalidations, how can a store be propagated to other non-local L1 caches? (I.e., how can the store be made visible to threads belonging to other CTAs?) Without writer-initiated invalidations, how can a consistency model—let alone a strong consistency model—be enforced?

In Sections 10.1.3 and 10.1.4, we discuss two proposals that employ self-invalidation [18], whereby a processor invalidates lines in its local cache to ensure that stores from other threads become visible.

What consistency model can the self-invalidation protocols enforce efficiently? We argue that such protocols are amenable to efficiently enforcing relaxed consistency models directly, rather than enforcing consistency-agnostic invariants such as the SWMR invariant.

Whether or not the consistency model should include scopes is under debate [15, 16, 23]. Whereas scopes adds complexity for programmers, they arguably simplify coherence implementations. Because scopes cannot be ruled out, we outline a scoped consistency model that does not limit synchronization to within a subset of scopes in Section 10.1.4. It is worth noting that the consistency model we introduce is similar in spirit to the ones used in today's industrial GPUs (which all use scoped models that do not limit synchronization to within a subset of scopes).

10.1.3 TEMPORAL COHERENCE

In this section, we discuss a self-invalidation based approach for enforcing coherence called temporal coherence [30]. The key idea is that each reader brings in a cache block for a finite period of time called the *lease*, at the end of which time the block is self-invalidated. We discuss two variants of temporal coherence: (1) a consistency-agnostic variant that enforces SWMR, in which a writer stalls until all of the leases for the block expire; and (2) a more efficient consistency-directed variant that directly enforces a relaxed consistency model, in which FENCEs, rather than writers, stall. For the following discussion, we assume that the shared cache (L2) is inclusive: a block not present in the shared L2 implies that it is not present in any of the local L1s. We also assume that the L1s use a write-through/no write-allocate policy: writes are directly written to the L2 (write-through) and a write to a block that is not present in the L1 does not allocate a block in the L1 (no write-allocate).

Consistency-agnostic Temporal Coherence

Instead of a writer invalidating all sharers in non-local caches, as is common with CPU coherence, consider a protocol in which the writer is made to wait until all of the sharers have evicted the block. By making the write wait until there are no sharers, the protocol ensures that there are no concurrent readers at the instant when the write succeeds—thereby enforcing SWMR.

How can the writer know how long to wait? That is, how can the writer ascertain that there are no more sharers for the block? Temporal coherence achieves this by leveraging a global notion of time. Specifically, it requires that each of the L1s and the L2 have access to a register that keeps track of global time.

On an L1 miss, the reader predicts how long it expects to hold the block in the L1, and informs the L2 of this time duration known as the *lease*. Every L1 cache block is tagged with a timestamp (TS) that holds the lease for that block. A read for an L1 block with current time greater than its lease is treated as a miss.

Furthermore, each block in the L2 is tagged with a timestamp. When a reader consults the L2 on an L1 cache miss, it informs the L2 of its desired lease; the L2 updates the timestamp for the block subject to the invariant that the timestamp holds the latest lease for that block across all L1s.

Every write—even if the block is present in the L1 cache—is written through to the L2; the write request accesses the block's timestamp held in the L2, and if the timestamp corresponds to a time in the future, the write stalls until this time. This stalling ensures that there are no sharers of the blocks when the write performs in the L2, thereby ensuring SWMR.

Example. In order to understand how temporal coherence works, let us consider the message passing example in Table 10.3, ignoring the FENCE instructions for now. Let us assume that threads T1 and T2 are from two different CTAs mapped to two different SMs (SM1 and SM2) with separate local L1s.

Table 10.3: Message passing example. T1 and T2 belong to different CTAs.

Thread T1	Thread T2	Comments
St1: St data1 = NEW;	Ld1: Ld r1 = flag;	// Initially all variables are 0.
St2: St data2 = NEW;	B1: if (r1 ≠ SET) goto Ld1;	
FENCE;	FENCE;	
St3: St flag = SET;	Ld2: Ld r2 = data2;	Can r2==0?

We illustrate a timeline of events at SM1, SM2, and the shared L2 in Table 10.4. Initially, let us assume that flag, data1, and data2 are cached in the local L1 of SM2 with lease values of 35, 30, and 20, respectively. At time=1, St1 is performed. Since the L1s use a write-through/no write-allocate policy, a write request is issued to the L2. Since data1 is valid in the L1 cache of SM2 until time=30, the write stalls until this time. At time=31, the write performs at the L2. St2 then issues a write request to the L2 at time=37 that reaches the L2 at time=42. By this time the lease for data2 (20) would have already expired, and so the write performs at the L2 without any stalling. Similarly, St3 issues a write request at time=48, and flag is written to

Table 10.4: Temporal coherence: Timeline of events for example in Table 10.3

Time	SM1	SM2	L2
/* T1 and T2 (belonging to different CTAs) are mapped to SM1 and SM2, respectively. flag = 0 (lease = 35), data1 = 0 (lease = 30) and data2 = 0 (lease = 20) are cached in L1 of SM2 */			
1	St1 issues write request		
6			write for data1 stalled until lease for data1 (30)
31			data1 = NEW written; Ack sent to SM1
36	St1 completes		
37	St2 issues write request		
42			data2 = NEW written without stalling since current time>lease for data2 (20); Ack sent to SM1
47	St2 completes		
48	St3 issues write request		
50		L1 lease for flag (35) has expired, so Ld1 issues read request	
53			flag = SET written without stalling since current time>lease for flag (35); Ack sent to SM1
55			flag = SET read; get new lease; sent to L1 of SM2
58	St3 completes		
60		Ld1 completes	
61		L1 lease for data2 (20) has expired, Ld2 issues read request	
66			data2 = NEW read; gets new lease; sent to L1 of SM2
71		Ld2 completes	

the L2 at time=53 without any stalling, since the lease for flag would have expired by this time. Meanwhile, at time=50, Ld1 checks its L1 and finds that the lease for flag has expired, so a read request is issued to the L2 and completes at time=60. Likewise, Ld2 also issues a read request to the L2 at time=61, reads the expected value of NEW from the L2 and completes at time=71. Because this variant of temporal coherence enforces SWMR, as long as GPU threads issue memory operations in program order, SC[5] is enforced.

Protocol Specification. We present the detailed protocol specification for the L1 and L2 controllers in Tables 10.5 and 10.6, respectively. The L1 has two stable states: (I)nvalid and (V)alid. The L1 communicates with the L2 using the *GetV(t)*, *Write*, and *WriteV(t)* requests. GetV(t), which carries a timestamp as a parameter, asks for the specified block to be brought into the L1 in valid state for the requested lease period, at the end of which time the block is self-invalidated. The Write request asks for the specified value to be written through to the L2 without bringing the block into the L1. The WriteV(t) request is used for writing to a block that is already valid in L1, and carries a timestamp holding its current lease as a parameter. For presentational reasons, we present only a high-level specification of the protocol (and for all other protocols in this chapter). Specifically, we show only stable states and the transitions between the stable states.

A load in state I causes a GetV(t) request to be sent to the L2; upon the receipt of data from the L2, the state transitions to V. A store in state I causes a Write request to be sent to the L2. Upon the receipt of an ack from the L2—indicating that the L2 has written the data—the store completes. Since the L1 uses a no-write-allocate policy, data is not brought to the L1 and the state of the block remains in state I.

Recall that a block becomes invalid if the global time advances past the lease for that block—this is represented logically by the V to I transition upon lease expiry.[6] Upon a store to a block in state V, a WriteV(t) request is sent to the L2. The purpose of WriteV(t) is to exploit the fact that if the block is held privately in the L1 of the writer, there is no need for the write to stall at the L2. Instead, the write can simply update the L2 as well as the L1, and continue to cache the block in the L1 until its lease runs out. The reason why the WriteV(t) request carries a timestamp is to determine whether or not the block is held privately in the L1 of the writer, as we will see next.

We now describe the L2 controller. The L2 has four stable states: (I)nvalid, indicating that the block is neither present in the L2 nor in any of the L1s; (P)rivate, indicating that the block is present in exactly one of the L1s; (S)hared, indicating that the block may be present in one or more L1s; and (Exp)ired, indicating that the block is present in the L2, but not valid in any of the L1s. Upon receiving a GetV(t) request in state I, the L2 fetches the block from

---

[5]The original paper [30] considered a warp as equivalent to a thread, whereas we consider a more modern GPU setting where threads can be individually scheduled. Consistency in our setting is therefore subtly different from their setting.

[6]It is worth noting that in the actual implementation it is not necessary for hardware to proactively detect that a block has expired and transition it to I state—the fact that a block has expired may be discovered lazily on a subsequent transaction to the block.

memory, updates the block's timestamp in accordance with the requested lease, and transitions to P. Upon receiving a Write in state I, it fetches the block from memory, updates the value in the L2, sends back an ack, and transitions to Exp, as it is not valid in any of the L1s.

Upon receiving a GetV(t) request in state P, the L2 responds with the data, extends the timestamp if the requested lease is greater than the current timestamp, and transitions to S. For a WriteV(t) request received in state P, there are two cases. In the straightforward case, the only SM holding the block privately writes to it; in this case, the L2 can simply update the

Table 10.5: Enforcing SWMR via Temporal Coherence: L1 Controller

	Load	Store	Eviction /Expiry
I	send GetV(t) to L2 rec. Data from L2 /V	send Write to L2 rec. Write-Ack from L2	
V	read hit	send WriteV(t) to L2 rec. Write-Ack from L2	-/I

Table 10.6: Enforcing SWMR via Temporal Coherence: L2 Controller

	GetV(t)	Write	WriteV(t)	L2: Eviction	L2: Expiry
I	send Fetch to Mem rec. Data from Mem send Data to L1 TS←t /P	send Fetch to Mem rec. Data from Mem write send Write-Ack to L1 /Exp			
P	send Data to L1 TS←max(TS,t) /S	stall (until expiry)	if(t = TS) write send Write-Ack to L1 else stall (until expiry)	stall (until expiry)	/Exp
S	send Data to L1 TS←max(TS,t)	stall (until expiry)	stall (until expiry)	stall (until expiry)	/Exp
Exp	send Data to L1 TS←t /P	write send Write-Ack to L1	write send Write-Ack to L1	if(dirty) send Write-back to Mem rec. Ack from Mem /I	

block without any stalling and reply with an ack. But there is a tricky corner case in which the WriteV(t) message from a private block is delayed such that the block is now held privately, but in a different SM! In order to disambiguate these two cases, every WriteV(t) message carries a timestamp with the lease for that block, and a match with the timestamp held in the L2 indicates the former straightforward case. A mismatch indicates the latter, in which case the WriteV(t) request stalls until the block expires, at which time the L2 is updated and an ack is sent back to the L1. A WriteV(t) or a Write request received in state S, on the other hand, will have to always wait until the block's lease at the L2 expires. Finally, an L2 block is allowed to be evicted only in Exp state because writes must know how long to stall in order to enforce SWMR.

In summary, we saw how temporal coherence enforces SWMR using leased reads and stalling writes. In combination with a processor that presents memory operations in program order, temporal coherence can be used to enforce sequential consistency (SC).

Consistency-directed Temporal Coherence

Temporal coherence as described previously enforces SWMR but at a significant cost, with every write to an unexpired shared block needing to stall at the L2 controller. Since the L2 is shared across all threads, stalling at the L2 can indirectly affect all of the threads, thereby reducing overall GPU throughput.

Recall that in Section 2.3 we discussed two classes of coherence interfaces: consistency-agnostic and consistency-directed. A consistency-agnostic coherence interface enforces SWMR by propagating writes to other threads synchronously before the write returns. Given the cost of enforcing SWMR in the GPU setting, can we explore consistency-directed coherence? That is, instead of making writes visible to other threads synchronously, can we make them visible asynchronously without violating consistency? Specifically, a relaxed consistency model such as Chapter 5's XC[7] mandates only the memory orderings indicated by the programmer via FENCE instructions. Such a model allows for writes to propagate to other threads asynchronously.

Considering the message-passing example shown in Table 10.3, XC merely requires that St1 and St2 become visible to T2 before St3 becomes visible. It does not mandate that by the time St1 performs it must have propagated to T2. In other words, XC does not require SWMR. Consequently, XC permits a variant of temporal coherence in which only FENCEs, rather than writes, require stalling. When a write request reaches the L2 and the block is shared, the L2 simply replies back to the thread initiating the write with the timestamp associated with the block. We refer to this time as the Global Write Completion Time (GWCT), as this indicates the time until which the thread must stall upon hitting a FENCE in order to ensure that the write has become globally visible to all threads.

For each thread mapped to an SM, the SM keeps track of the maximum of GWCTs returned for the writes in the per-thread *stall-time* register. Upon hitting a FENCE, stalling the

---

[7]For now, let us assume there are no scopes in the memory model.

thread until this time ensures that all of the writes before the FENCE have become globally visible.

Example. We illustrate a timeline of events in Table 10.7 for the same message-passing example (Table 10.3), but now taking into account the effect of FENCE instructions, because we are seeking to enforce a relaxed XC-like model.

At time=1, a write request due to St1 is issued to the L2. At time=6, the write is performed at the L2 without any stalling although its lease (30) would not have expired then; L2 replies with a GWCT of 30 which is remembered at SM1 in its per-thread stall-time register. In a similar vein, a write request due to St2 is issued at time=12. At time=17, the write is performed at the L2 and the L2 replies back with a GWCT of 20. Upon receiving the GWCT, SM1 does not update its stall-time because the current value (30) is higher. The FENCE instruction is executed at time=23 and blocks thread T1 until its stall-time of 30. At time=31, a write-request due to St3 is issued to the L2 and the write is performed at the L2 at time=36. Because the lease for flag (35) would have expired, the L2 does not respond with a GWCT, and St3 completes at time=41. Meanwhile, at time=40, Ld1 attempts to read flag from the L1 but its lease would have expired at time=35. Therefore, a read request for flag is issued to the L2, reads SET from the L2 at time=45, and completes at time=50. In a similar vein, a read request due to Ld2 is issued at time=51, reads from L2 at time=56 and completes at time=57, returning the expected value of NEW.

Protocol Specification. The consistency-directed temporal coherence protocol specifications are mostly similar to the consistency-agnostic variant—we highlight the differences in bold in Table 10.8 (L1 controller) and Table 10.9 (L2 controller).

The main difference with the L1 controller (Table 10.8) is due to the fact that Write-Acks from the L2 now carry GWCTs. Accordingly, upon receiving a Write-Ack, the L1 controller extends the stall-time if the incoming GWCT is greater than the currently held stall-time for that thread. (Recall that upon hitting a FENCE, the thread is stalled until the time held in the stall-time register.)

The main difference with the L2 controller (Table 10.9) is that Write requests do not induce a stall; instead, the write is performed and a GWCT is returned along with the Write-Ack. In a similar vein, a WriteV(t) request in state S also does not stall.

Temporal Coherence: Summary and Limitations

We saw how temporal coherence can either be used to enforce SWMR or directly enforce a relaxed consistency model such as XC.[8] The key benefit with the latter approach is that it eliminates expensive stalling at the L2; instead writes stall at the SM upon hitting a FENCE. More optimizations are possible that further reduce stalling [30]. However, there are some critical limitations with temporal coherence.

---

[8]Temporal coherence [30] enforces a variant of XC in which writes are not atomic.

Table 10.7: Consistency-directed Temporal coherence: Timeline for Table 10.3

Time	SM1	SM2	L2
	/* T1 and T2 (belonging to different CTAs) are mapped to SM1 and SM2, respectively. flag = 0 (lease = 35), data1 = 0 (lease = 30), and data2 = 0 (lease = 20) are cached in L1 of SM2 */		
1	St1 issues write request		
6			data1=NEW written although current time < lease for data1 (30); Ack sent to SM1 with GWCT=30
11	St1 completes, stall-time ←30		
12	St2 issues write request		
17			data2=NEW written although current time < lease for data2 (20); Ack sent to SM1 with GWCT=20
22	St2 completes; GWCT (20)< stall-time(30), so stall-time unchanged		
23	FENCE stalls thread until stall-time (30)		
31	St3 issues write request		
36			flag=SET written since current time > lease for flag (35); Ack sent to SM1
40		L1 lease for flag (35) has expired, so Ld1 issues read request	
41	St3 completes		
45			flag=SET read; gets new lease; sent to L1 of SM2
50		Ld1 completes	
51		L1 lease for data2 (20) has expired so Ld1 issues read request	
56			data2=NEW read; gets new lease; sent to L1 of SM2
57		Ld2 completes	

Table 10.8: Consistency-directed Temporal Coherence: L1 Controller (diffs in bold)

	Load	Store	Eviction/ Expiry
I	send GetV(t) to L2 rec. Data from L2 /V	send Write to L2 rec. Write-Ack+**GWCT** from L2 **stall-time←max(stall-time,GWCT)**	
V	read hit	send WriteV(t) to L2 rec. Write-Ack+**GWCT** from L2 **stall-time←max(stall-time,GWCT)**	-/I

Table 10.9: Consistency-directed Temporal Coherence: L2 Controller (diffs in bold)

	GetV(t)	Write	WriteV(t)	L2: Eviction	L2: Expiry
I	send Fetch to Mem rec. Data from Mem send Data to L1 TS←t /P	send Fetch to Mem rec. Data from Mem Write send Write-Ack to L1 /Exp			

P	send Data to L1
	TS←max(TS,t)
	/S
		~~stall~~
		write
		send Write-Ack+GWCT to L1
			if(t = TS)
			    write
			    send Write-Ack to L1
			else
			    stall (until expiry)
				stall
				(until expiry)
					/Exp

S	send Data to L1 TS←max(TS,t)	~~stall~~ **write** **send Write-Ack+GWCT to L1**	~~stall~~ **write** **send Write-Ack+GWCT to L1**	stall (until expiry)	/Exp

Exp	send Data to L1
	TS←t
	/P
		write
		send Write-Ack to L1
			write
			send Write-Ack to L1
				if(dirty)
				    send Write-back to Mem
				    rec. Ack from Mem
				/I	

- Supporting a non-inclusive L2 cache is cumbersome. This is because temporal coherence requires that a block that is valid in one or more L1s must have its lease time available at the L2. A complex workaround is possible wherein an unexpired block may be evicted from the L2 provided the evicted block's lease is held someplace, for instance along with the L2 *miss status holding register* (MSHR) [30].

- It requires global timestamps. With modern GPUs being relatively large area-wise, maintaining globally synchronized timestamps could be hard. A recent proposal, however, has shown how a variant of temporal coherence can be implemented without using global timestamps [25].

- Performance could be sensitive to the choice of the lease period. A lease period that is too short increases the L1 miss rate; a lease period that is too long causes the writes (or FENCEs) to stall more.

- Temporal coherence cannot directly take advantage of scoped synchronization. For instance, stores involved in CTA-scoped synchronization (intuitively) need not be written to the L2s—but in temporal coherence every store is written to the L2 since it is designed under the assumption of write-through/no-write-allocate L1 caches.

- Temporal coherence involves timestamps. Timestamps introduce complexities (e.g., timestamp rollover) in the design and verification process.

In summary, although workarounds are possible for most of the above limitations, they do tend to add complexity to an already unconventional timestamp-based coherence protocol.

10.1.4   RELEASE CONSISTENCY-DIRECTED COHERENCE

Temporal coherence, which is based on the powerful idea of leases, is versatile enough to enforce both consistency-agnostic and consistency-directed variants of coherence. On the other hand, protocols involving leases and timestamps are arguably cumbersome. In this section, we discuss an alternative approach to GPGPU coherence called *release consistency-directed coherence* (RCC) that directly enforces release consistency (RC), which differs from XC by distinguishing acquires from releases, whereas XC treats all synchronization the same.

RCC compromises on flexibility, in that it can only enforce variants of RC. But in return for this reduced flexibility, RCC is arguably simpler, can naturally exploit scope information, and can be made to work with non-inclusive L2 caches. In the following, we start by briefly recapping the RC memory model and then extend it with scopes. We then describe a simple RCC protocol followed by two optimizations. Each of the protocols can enforce both the scoped and non-scoped variants of RC.

Release Consistency: Non-scoped and scoped variants

In this section, we discuss the RC memory model starting with the non-scoped variant, and then extend it with scopes.

Recall that RC (introduced in Section 5.5.1) has special atomic operations that order memory accesses in one direction as opposed to the bidirectional ordering enforced by a FENCE. Specifically, RC has a release (**Rel**) store and an acquire (**Acq**) load that enforce the following orderings.

- **Acq** Load → Load/Store

- Load/Store → **Rel** Store

- **Rel** Store/**Acq** Load → **Rel** Store/**Acq** Load

Consider the message-passing example shown in Table 10.10. Marking St2 as a release ensures St1→St2; marking Ld1 as an acquire ensures Ld1→Ld2 orderings. The acquire[9] (Ld1) reads the value written by the release (St2). In doing so, the release synchronizes with the acquire, ensuring St2→Ld1 in the global memory order. Combining all of the above orderings implies St1→Ld2, thus ensuring that Ld2 sees the new value and not 0.

Table 10.10: Message passing example. Non-scoped RC.

Thread T1	Thread T2	Comments
St1: St data1 = NEW; St2: **Rel** St flag = SET;	Ld1: **Acq** Ld r1 = flag; B1: if (r1 ≠ SET) goto Ld1; Ld2: Ld r2 = data1	// Initially all variables are 0. Can r2==0?

In the variant of the memory model without scopes, a release synchronizes with an acquire as long as the acquire returns the value written by the release, irrespective of whether the threads to which they belong are from the same scope or different scope. Thus, in the above example, Ld2 will see the new value irrespective of whether T1 and T2 belong to the same CTA or different CTAs.

A Scoped RC Model.

In a scoped RC model, each atomic operation is associated with a scope. A release is said to synchronize with an acquire if: (1) the acquire load returns the value written by the release store and (2) the scope of each atomic operation includes the thread executing the other operation.

For instance, a release of CTA scope is said to synchronize with an acquire of CTA scope only if the two threads executing the acquire and release belong to the same CTA. On the other hand, a release of GPU scope is said to synchronize with an acquire of GPU scope irrespective

[9]Acquire refers to the load marked acquire and release refers to a store marked release.

of whether the two threads are from the same CTA or different CTAs (as long as they are issued on the same GPU).

Table 10.11: Message passing example. Scoped RC.

Thread T1	Thread T2	Comments
St1: St data1 = NEW;	Ld1: **CTA Acq** Ld r1 = flag;	// Initially all variables are 0.
St2: **GPU Rel** St flag = SET;	B1: if (r1 ≠ SET) goto Ld1;	
	Ld2: Ld r2 = data1	Can r2==0? (could be 0, if T1 and T2 are from different CTAs)

For more intuition, consider the scoped variant of the message-passing example shown in Table 10.11. As we can see, the release St2 carries a GPU scope whereas the acquire Ld1 carries only a CTA scope. If T1 and T2 are from different CTAs, then the scope of the acquire (Ld1) does not include T1. Therefore, in such a situation the release is said not to synchronize with the acquire, which means that r2 could in fact read the old value of 0. On the other hand, if T1 and T2 are from the same CTA, r2 cannot read a 0.

One approach to formalizing scoped RC is to employ a variant of Shasha and Snir's formalism [28] and use a partial order, instead of a global memory order, to order conflicting operations. More specifically, not all conflicting operations are ordered—only releases and acquires that synchronize with each other are ordered. It is worth noting that NVIDIA adopted this approach to formalizing its PTX memory consistency model [20].

Release Consistency-directed Coherence (RCC)

In this section, we introduce a protocol that directly enforces RC instead of enforcing SWMR. Specifically, the protocol does not eagerly propagate writes to other threads—i.e., it does not enforce SWMR. Instead, writes are written to the L2 upon a release and become visible to another thread when that thread self-invalidates the L1 on an acquire and pulls in new values from the L2.

For the following discussion let us assume write-back/write-allocate L1 caches. Also, for now let us ignore scopes and assume that synchronization is between threads from two different CTAs. We will describe later how RCC can handle intra-CTA synchronization efficiently. The main steps involved in RCC are as follows.

- Loads and stores that are not marked acquire or release behave like normal loads and stores in a write-back/write-allocate L1 cache.

Table 10.12: RCC: L1 controller

	Load/ Acq.scope = CTA	Store/ Rel.scope = CTA	Acq.scope = GPU/ Acq (no scope)	Rel.scope = GPU/ Rel (no scope)	L1: Eviction
I	send GetV to L2 rec. Data from L2 read/V	send GetV to L2 rec. Data from L2 write/V	send GetV to L2 rec. Data from L2 read/V forall dirty blocks send Write-back to L2 rec. Ack from L2 forall other valid blocks Invalidate	forall dirty blocks send Write-back to L2 rec. Ack send GetV to L2 rec. Data from L2 write/V send Write-back to L2 rec. Ack	
V	read hit	write hit	send GetV to L2 rec. Data from L2 read forall dirty blocks send Write-back to L2 rec. Ack from L2 forall other valid blocks Invalidate	forall dirty blocks send Write-back to L2 rec. Ack write hit send Write-back to L2 rec. Ack	if(dirty) send Write-back to L2 rec. Ack from L2 /I

Table 10.13: RCC: L2 controller

	GetV	Write-back	L2: Eviction
I	send Fetch to Mem rec. Data from Mem send Data to L1 /V	allocate block write send Ack to L1 /V	
V	send Data to L1	write send Ack to L1	if(dirty) send Write-back to Mem rec. Ack from Mem /I

- Upon a store marked release, all dirty blocks in the L1, except the one written by the release, are written to the L2. Then, the block written by the release is written to the L2, ensuring Load/Store → **Rel** Store.

- A load marked acquire reads a fresh copy of the block from the L2. Then, all valid blocks in the L1, other than the one read by the acquire, are self-invalidated, thereby ensuring **Acq** Load → Load/Store.

Table 10.14: RCC: Timeline of events (for Table 10.10)

Time	SM1/L1	SM2/L1	L2
	/* T1 and T2 (belonging to different CTAs) are mapped to SM1 and SM2, respectively. flag = 0 and data1 = 0 are cached in L1s of both SM1 and SM2 */		
t_1	St1 writes data1=NEW and completes		
t_2	**Rel** St2 issues Write-back of data1		
t_3			data1=NEW written and Ack sent back to L1 of SM1
t_4	Ack received; flag=SET written and its Write-back initiated		
t_5			flag=SET written and Ack sent back to L1 of SM1
t_6	**Rel** St2 completes		
t_7		**Acq** Ld1 issues read for flag	
t_8			flag=SET read and sent to L1 of SM2
t_9		value received; data1 self-invalidated; **Acq** Ld1 completes	
t_{10}		Ld2 issues read for data1	
t_{11}			data1=NE W read and sent to L1 of SM2
t_{12}		Ld2 completes	

Example. Consider the message-passing example in Table 10.10, assuming that T1 and T2 are from different CTAs, mapped to two different SMs (SM1 and SM2). Initially let us assume the cache blocks containing data1 and flag are cached in the L1s of both SM1 and SM2. We illustrate a timeline of events in Table 10.14. At time=t_1, St1 is performed, causing NEW to be written to data1 cached in SM1's L1. At time=t_2, St2 (marked release) causes all of the dirty blocks in SM1's L1 to be written to the L2. Therefore, data1 is written to the L2 at time=t_3. Then, at time=t_4, the release write is performed, causing SET to be written to flag cached in SM1's L1 and subsequently written back to the L2 at time=t_5. At time=t_7, Ld1 (marked acquire) issues a read for flag, causing flag=SET to be read from the L2. At time=t_9, when the value is received, all of the valid blocks in SM2's L1, except the block read by the acquire, are self-invalidated.

Thus, data1 is self-invalidated. At time=t_{10}, Ld2 causes a read request for data1 to be sent to the L2 and the up-to-date value of NEW is read.

Exploiting Scopes. If all atomic operations are of GPU scope, the protocol described above is applicable as is. This is because the protocol already pushes all of the dirty data to the L2 (the cache level corresponding to GPU scope) on a release, and pulls data from the L2 upon an acquire. Notably, the RCC protocol can take advantage of CTA scopes: a CTA-scoped release need not write-back dirty blocks to the L2; a CTA-scoped acquire need not self-invalidate any of the valid blocks in the L1.

Protocol Specification. We present the detailed specifications of the L1 and L2 controllers in Tables 10.12 and 10.13. There are two stable states: (I)nvalid and V(alid). We assume every byte within a block is tagged with a dirty bit; if any of the bytes is tagged dirty, the block as a whole is also tagged dirty. The L1 communicates with the L2 using the *GetV* and the *Write-back* requests. GetV asks for the specified block to be brought into the L1 in Valid state. Differently from GetV(t) used in temporal coherence, GetV does not carry a timestamp as a parameter; a block brought into the L1 via GetV is self-invalidated upon an acquire. The Write-back request copies the dirty bytes within the specified block from the L1 into the L2, without evicting the block from the L1.

Table 10.12 shows the L1 controller. Releases and acquires of CTA scope behave like ordinary stores and loads in a write-back/write-allocate cache. If the release and acquire are of GPU scope (or if the releases and acquire carry no scope information), the protocol must write-back all dirty blocks on a release and self-invalidate all valid blocks on an acquire. Before the blocks are self-invalidated on an acquire, dirty blocks, if any, are written to the L2 to ensure that their values are not lost. (Alternatively, it is permissible to delay writing back the dirty blocks until the next release.)

Finally, it is worth noting that the protocol does not rely on cache inclusion. Indeed, as shown in Table 10.13, a valid L2 block can be silently evicted without informing the L1. Intuitively, this is because the L2 does not hold any critical metadata such as sharers or ownership information.

Summary. In summary, RCC is a simple protocol that directly enforces RC. Because it does not hold any protocol metadata at the L2, it does not require the L2 to be inclusive and allows for L2 blocks to be evicted silently. The protocol can take advantage of scope information—specifically, if the release and acquire are of CTA scope it does not require expensive write-backs or self-invalidations. On the other hand, without knowledge of scopes, intra-CTA synchronization is inefficient: RCC has to assume conservatively that releases and acquires are of GPU scope and write-back/self-invalidate data from the L1 even if the synchronization is within one CTA.

Exploiting Ownership: RCC-O

Releases (which cause *all* dirty lines to be written back to the L2) and acquires (which cause *all* valid lines to be self-invalidated) are expensive operations in the simple protocol described

Table 10.15: RCC-O: L1 controller. In non-scoped version, scope=GPU.

	Load/ Acq.scope = CTA	Store/ Rel.scope = CTA	Acq.scope = GPU/ Acq (no scope)	Rel.scope = GPU/ Rel (no scope)	L1: Eviction	From L2: Req-Write-back
I	send GetV to L2 rec. Data from L2 read/V	send GetO to L2 rec. Data from L2 write/O	send GetV to L2 rec. Data from L2 read/V forall other valid non-O blocks Invalidate	send GetO to L2 rec. Data from L2 write/O		
V	read hit	send GetO to L2 rec. Data from L2 write/O	send GetV to L2 rec. Data from L2 read forall other valid non-O blocks Invalidate	send GetO to L2 rec. Data from L2 write/O	/I	
O	read hit	write hit	read hit	write hit	send Write-back to L2 rec. Ack from L2/I	send Data to L2/V

above. One approach to reduce the cost of releases and acquires is to track *ownership* [13, 16], so that owned blocks need not be self-invalidated on an acquire nor written-back upon a release.

The key idea is to add an O(wned) state—every store must obtain ownership of the L1 cache line before a subsequent release. For each block, the L2 maintains the owner, which refers to the identity of the L1 that owns the block. Upon a request for ownership, the L2 first downgrades the previous owner (if there is one), causing the previous owner to write-back dirty data to the L2. After sending the current data to the new owner, the L2 changes ownership. Because a block in state O implies the absence of remote writers, there is no need to self-invalidate an Owned block on an acquire. Because the L2 tracks owners, there is no need to write-back an Owned block upon a release.

There is yet another benefit to ownership tracking: even in the absence of scope information, ownership tracking can help in reducing the cost of intra-CTA acquires. Consider the message passing example shown in Table 10.10, assuming now that the threads T1 and T2 belong to the same CTA. In the absence of scope information, recall that RCC will have to treat both releases and acquires conservatively, writing back all of the dirty data on a release, and self-invalidating all valid blocks on an acquire. With RCC-O, releases still have to be treated conservatively; all of the stores before the release will still have to obtain ownership. If a block that is acquired is in Owned state, it implies that the corresponding release must have been from a thread that is within the same CTA. Consequently, the acquire can be treated like a CTA-scoped acquire and the self-invalidation of valid blocks can be obviated.

Protocol Specification. We present the detailed specification of the L1 controller in Table 10.15 and the L2 controller in Table 10.16. The main change is the addition of the O state. Every store in V (or I) state has to contact the L2 controller and request ownership. Having obtained own-

Table 10.16: **RCC-O: L2 controller**

	GetV	GetO	Write-back	L2: Eviction
I	send Fetch to Mem rec. Data from Mem send Data to L1 /V	send Fetch to Mem rec. Data from Mem send Data to L1 set owner/O	allocate block write send Ack to L1/V	
V	send Data to L1	send Data to L1 set owner/O	write send Ack to L1	/I
O	send Req-Write-back to L1 (owner) rec. Data from L1 (owner) write/V send Data to L1 (requestor)	send Req-Write-back to L1 (owner) rec. Data from L1 (owner) send Data to L1 (requestor) update owner/O	write send Ack to L1/V	send Req-Write-back to L1 (owner) rec. Data from L1 (owner) send Write-back to Mem rec. Ack from Mem /I

ership, any subsequent store to that line can simply update the line in the L1 without contacting the L2. Thus, a block in O state is potentially stale at the L2. For this reason, when the L2 receives a request for ownership of a line already in O state, the L2 first requests the previous owner to write-back the block, sends the up-to-date data to the current requester, and then changes ownership.

Upon hitting a release (CTA-scoped, GPU-scoped, or non-scoped), there is no need to write-back dirty data from the L1 to the L2. Instead, the protocol merely requires that all previous stores must have obtained ownership. Upon an acquire (GPU-scoped or non-scoped), only non-owned valid blocks need to be self-invalidated. If the acquire is of CTA scope or if the acquire is to a block that is already owned, it implies that the synchronization is intra-CTA and hence there is no need for self-invalidation.

Finally, because the protocol hinges on the L2 maintaining ownership information, an L2 block in O state cannot be silently evicted; instead, it must first downgrade the current owner, requesting it to write-back the block in question.

Summary. By tracking ownership, RCC-O allows for a reduced number of self-invalidations in comparison to RCC. In the absence of scopes, ownership tracking also enables the detection of intra-CTA acquires and obviates the need for self-invalidation in such a situation. However, intra-CTA releases are treated conservatively—each store to a previously unowned block still obtains ownership before a subsequent release. Finally, RCC-O cannot allow for blocks in state O to be evicted from the L2 silently; it has to first downgrade the current owner.

Lazy Release Consistency-directed Coherence: LRCC
In the absence of scope information, RCC treats both releases and acquires conservatively. In RCC-O, intra-CTA acquires can be detected and efficiently handled but releases are treated conservatively.

Is there a way for intra-CTA releases to also be efficiently handled? Lazy release consistency-directed coherence (LRCC) enables this. In LRCC, ordinary stores do not obtain ownership. Only release stores obtain ownership, on behalf of the stores before it in program order. When a block that is released is later acquired, the state of the block at the time of the acquire determines the coherence actions.

If the block is owned by the L1 of the acquiring thread, it implies that the synchronization is intra-CTA; hence, there is no need for self-invalidation. If the block instead is owned by a remote L1, it implies that the synchronization is inter-CTA. In that case, dirty cache blocks in that remote L1 are first written back and the valid blocks in the acquiring L1 are self-invalidated. Thus, by delaying coherency actions until an acquire, LRCC is able to handle intra-CTA synchronization very efficiently.

Example: Inter-CTA synchronization. Consider the message passing example shown in Table 10.10, assuming that the threads T1 and T2 belong to different CTAs. Initially, let us assume the cache blocks containing data1 and flag are cached in the L1s of both SM1 and SM2. We illustrate a timeline of events in Table 10.17. At time=t_1, St1 is performed causing NEW to be written to data1 cached in SM1's L1, without obtaining ownership. At time=t_2, St2 (marked release) is performed, causing an ownership request for flag to be sent to the L2. With flag previously unowned by any SM, the L2 grants ownership by simply setting the owner to SM1 and responds to SM1. Upon receiving the response, SET is written to flag at SM1's L1, and St2 completes at time=t_4. At time=t_5, Ld1 (marked acquire) is performed, causing a read request for flag to be sent to the L2. At the L2, the block containing flag is found to be owned by SM1's L1; therefore, the L2 requests for SM1's L1 to write-back flag. Upon receiving this request at time=t_7, SM1's L1 first writes back all dirty non-owned blocks (including data1=NEW); then, at time=t_9, it writes back flag=SET. At time=t_{10}, L2 receives flag, writes it locally and forwards the value to SM2. At time=t_{11}, SM2 receives flag=SET and Ld1 completes. Finally, at time=t_{12}, Ld2 causes a read request for data1 to be sent to the L2 and the up-to-date value of NEW is read.

Example: Intra-CTA synchronization. Let us consider the message passing example shown in Table 10.10, now assuming that the threads T1 and T2 belong to the same CTA, and hence are mapped to the same SM, SM1. We illustrate a timeline of events in Table 10.18. At time=t_1, St1 is performed, causing NEW to be written to data1 cached in the L1. At time=t_2, St2 (marked release) is performed, causing an ownership request for flag to be sent to the L2. Upon obtaining ownership, SET is written to flag cached in the L1. At time=t_5, Ld1 (marked acquire) is performed. Since flag is owned by SM1, there are no self-invalidations and flag=SET is simply read from the L1. At time=t_6, Ld2 is performed, reading data1=NEW from the L1.

Protocol Specification. We present the detailed specification of the L1 controller in Table 10.19. The L2 controller is identical to the one for RCC-O (Table 10.16). In contrast to RCC-O, not all stores obtain ownership; only non-scoped/GPU-scoped releases obtain own-

Table 10.17: LRCC: Timeline of events (for Table 10.10), inter-CTA synchronization

Time	SM1/L1	SM2/L1	L2
	/* T1 and T2 (belonging to different CTAs) are mapped to SM1 and SM2, respectively. flag = 0 and data1 = 0 are cached in L1s of both SM1 and SM2 */		
t_1	St1 writes data1=NEW and completes		
t_2	**Rel** St2 issues ownership request for flag		
t_3			ownership for flag obtained; and sent to SM1
t_4	ownership received; flag=SET written and St2 completes		
t_5		**Acq** Ld1 issues read for flag	
t_6			write-back request for flag issued to L1 of SM1 (since it owns flag)
t_7	(before responding with flag) write-back of data1 initiated		
t_8			data1=NEW written and Ack sent to L1 of SM1
t_9	Ack received; flag=SET sent to L2		
t_{10}			flag value written and sent to L1 of SM2
t_{11}		**Acq** Ld1 completes	
t_{12}		Ld2 issues read for data1	
t_{13}			data1=NEW read; sent to L1 of SM2
t_{14}		Ld2 completes	

ership. Like in RCC-O, a CTA-scoped acquire or an acquire for a block owned at the L1 implies intra-CTA synchronization; hence, there are no self-invalidations. Conversely, a GPU-scoped/non-scoped acquire for a block not owned by the L1 implies inter-CTA synchronization and involves self-invalidations. Specifically, a GetV request is sent to the L2 to obtain the up-to-date value of the block that is acquired. (If at the L2, the requested block is owned by a different L1, the L2 causes that L1 to write back all of its dirty, non-owned blocks before

Table 10.18: LRCC: Timeline of events (for Table 10.10), intra-CTA synchronization

Time	T1 (SM1/L1)	T2 (SM1/L1)	L2
/* Both T1 and T2, belonging to the same CTAs, are mapped to SM1. flag = 0 and data1 = 0 are cached in SM1's L1. */			
t_1	St1 writes data1=NEW and completes		
t_2	**Rel** St2 issues ownership request		
t_3			ownership for flag obtained and sent to SM1
t_4	ownership received; flag=SET written and St2 completes		
t_5		**Acq** Ld1 reads flag=SET from L1 and completes	
t_6		Ld2 reads data1=NEW from L1 and completes	

writing back the requested block.) Then all valid non-owned L1 blocks, except the block just read by the acquire, are self-invalidated. Before the blocks are self-invalidated, non-owned dirty blocks, if any, are written back to the L2 to ensure that their values are not lost. (Alternatively, it is permissible to delay writing back the dirty blocks until the next release.) Finally, like in RCC-O, LRCC hinges on the L2 maintaining ownership information. Therefore, an L2 block in state O cannot be evicted silently; instead, it must downgrade the current owner, asking it to write-back not only the block in question but also any other non-owned dirty blocks in that L1.

Summary. By delaying coherence actions until the acquire, LRCC allows for intra-CTA synchronization to be detected and handled efficiently even in the absence of scope information. LRCC cannot allow for blocks in state O to be silently evicted from the L2, but the impact of this is limited to synchronization objects, since only release stores obtain ownership.

Release Consistency-directed Coherence: Summary

In this section, we discussed three variants of release consistency-directed coherence that can be used to enforce both scoped and non-scoped flavors of release consistency. While all of the variants can exploit knowledge of scopes, LRCC can handle intra-CTA synchronization efficiently even in the absence of scopes.

Given that LRCC can handle synchronization efficiently even in the absence of scopes, can we get rid of scoped memory consistency models? Not quite, in our opinion. Whereas lazi-

Table 10.19: LRCC: L1 controller. In non-scoped version scope is set to GPU.

	Load/ Acq.scope = CTA	Store/ Rel.scope = CTA	Acq.scope = GPU/ Acq (no scope)	Rel.scope = GPU/ Rel (no scope)	L1: Eviction	From L2: Req-Write-back
I	send GetV to L2 rec. Data from L2 read/V	send GetV to L2 rec. Data from L2 write/V	send GetV to L2 rec. Data from L2 read/V forall dirty non-O blocks send Write-back to L2 rec. Ack from L2 forall other valid non-O blocks Invalidate	send GetO to L2 rec. Data from L2 write/O		
V	read hit	write hit	send GetV to L2 rec. Data from L2 read forall dirty non-O blocks send Write-back to L2 rec. Ack from L2 forall other valid non-O blocks Invalidate	send GetO to L2 rec. Data from L2 write/O	if(dirty) send Write-back to L2 rec. Ack from L2 /I	
O	read hit	write hit	read hit	write hit	for all other dirty non-O blocks send Write-back to L2 rec. Ack from L2/V send Write-back to L2 rec. Ack from L2/I	for all other dirty non-O blocks send Write-back to L2 rec. Ack from L2/V send Data to L2/V

ness helps in avoiding wasteful write-backs in case of intra-CTA synchronization, it makes inter-CTA acquires slower. This is because laziness forces all of the write-backs from the L1 of the releasing thread to be performed on the critical path of the acquire. This effect becomes more pronounced in the case of synchronization between two devices, e.g., between a CPU and a GPU.

Thus, whether or not GPU and heterogeneous memory models must involve scopes is a complex programmability vs. performance tradeoff.

10.2 MORE HETEROGENEITY THAN JUST GPUS

In this section, we consider the problem of how to expose a global shared memory interface across multiple multicore devices consisting of CPUs, GPUs, and other accelerators.

What makes the problem challenging is that each of the devices might guarantee a distinct consistency model enforced via a distinct coherence protocol. For instance, a CPU may choose to enforce a relatively strong consistency model such as TSO, using a consistency-agnostic protocol that enforces SWMR. On the other hand, a GPU may choose to enforce scoped RC, by employing a consistency-directed lazy release consistency protocol.

When two or more devices with distinct consistency models are integrated, what is the resulting consistency model of the heterogeneous device? How can it be programmed? How to integrate the coherence protocols of the two devices? Most of these questions are being actively researched by academia and industry. In the following section, we attempt to understand these questions better and outline the design space.

10.2.1 HETEROGENEOUS CONSISTENCY MODELS

Let us first understand the semantics of composing two distinct consistency models. Suppose two component multicore devices A and B are integrated such that they share memory—what is the resulting consistency model?

A seemingly straightforward answer is to declare the weaker of the two memory models to be the overall model. However, this simple solution will not work if the two memory models are incomparable as we saw in Section 5.8.1. Even if one of the models subsumes the other, a more efficient solution is possible.

For example, suppose multicore A enforces SC and multicore B satisfies TSO, how does the heterogeneous multicore made up of A and B behave?

Figure 10.2 shows the operational model of the heterogeneous multicore processor. Intuitively, operations coming out of a thread from A should satisfy the memory ordering rules of A (which in our example is SC) and operations coming out of B should satisfy the memory model ordering rules of B (TSO).

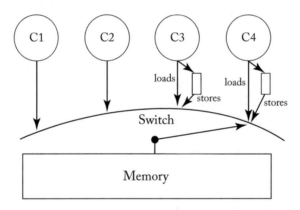

Figure 10.2: Semantics of a heterogeneous architecture in which cores C1 and C2 support SC, whereas C3 and C4 support TSO.

Consider the Dekker's example in Table 10.20. Is it possible for both Ld1 and Ld2 to read zeroes? This is indeed possible since TSO does not enforce the St2→Ld2 ordering. As shown in Table 10.21, once a FENCE instruction is inserted between St2 and Ld2, this is rendered impossible, since the FENCE now enforces the St2→Ld2 ordering. Note, however, that a FENCE is not needed between St1 and Ld1 since SC already enforces St1→Ld1.

Thus, a heterogeneous shared memory architecture that is formed by composing component multicores with distinct consistency models results in a compound consistency model such that memory operations originating from each component satisfy the memory ordering rules of that component. Although this notion seems intuitive, to our knowledge this has not been formalized yet.

Table 10.20: Dekker's example. Combined SC + TSO memory model. Thread T1 is mapped to an SC core whereas T2 is mapped to a TSO core.

Thread T1 (SC)	Thread T2 (TSO)	Comments

St1: St1 flag1 = NEW;
Ld1: Ld1 r1 = flag2;
	St2: St flag2 = NEW;
	Ld2: Ld r2 = flag1;
		// Initially all variables are 0.
		// Can both r1 and r2 read 0?
		// Yes!

Table 10.21: Dekker's example with a FENCE inserted on the TSO side. Combined SC + TSO memory model.

Thread T1 (SC)	Thread T2 (TSO)	Comments

St1: St flag1 = NEW;
Ld1: Ld   r1 = flag2;
	St2: St flag2 = NEW;
	FENCE
	Ld2: Ld r2 = flag1;
		// Initially all variables are 0.
		// Can both r1 and r2 read 0?
		// No!

Programming with Heterogeneous Consistency Models

How do we program such a heterogeneous shared memory architecture in which not all devices support the same consistency model? As we saw above, one approach is to simply target the compound consistency model. However, as one can imagine, programming with compound memory models can quickly get tricky, especially if very different memory consistency models are involved.

A more promising approach is to program with languages such as HSA or OpenCL with formally specified (scoped) synchronization primitives. Recall that C11 is a language model based on Sequential Consistency for data-race-free ("SC for DRF") that we saw in Section 5.4. HSA and OpenCL extend the SC for DRF paradigm with scopes, dubbed Sequential Consistency for Heterogeneous-Race-Free (SC for HRF) [15].

HRF is conceptually similar to the scoped RC model we introduced in the previous section, but with a couple of differences. First, HRF is at the language-level, whereas the scoped RC model is at the virtual ISA level. Second, whereas HRF does not provide any semantics for programs with races, the scoped RC model provides semantics for such codes. Thus, the relationship between HRF and scoped RC is akin to the relationship between DRF-based language models and RC.

HRF has two variants: HRF-direct and HRF-indirect. In HRF-direct, two threads can synchronize only if the synchronization operations (releases and acquires) have the same exact scope. HRF-indirect, on the other hand, allows for transitive synchronization using different scopes. For example, in HRF-indirect, if thread T1 synchronizes with T2 using scope S1, and subsequently T2 synchronizes with T3 using scope S2, T1 is said to transitively synchronize with T3. HRF-direct, which does not allow this transitivity, could be restrictive for programmers. It is worth noting that the scoped RC model, like HRF-indirect, allows for transitive synchronization.

The language level scoped-synchronization primitives are mapped to hardware instructions; each device with a unique hardware consistency model has separate mappings. Lustig et al. [21] provide a framework dubbed ArMOR for algorithmically deriving correct mappings based on their precise way to specify different memory consistency models.

The third approach toward programming with heterogeneity is to program with one of the component hardware memory consistency models, while instrumenting the code running in each of the other components with FENCE instructions and other instructions to ensure they are compatible with the chosen memory model. Again, the ArMOR framework can be used to translate between memory consistency models.

10.2.2 HETEROGENEOUS COHERENCE PROTOCOLS

Consider two multicore devices A and B with each device adhering to a distinct consistency model enforced via a distinct coherence protocol. As we can see in Figure 10.3, each of the devices has a local coherence protocol for keeping its local L1s coherent. How do we integrate the two devices into one heterogeneous shared memory machine? In particular, how do we stitch the two coherence protocols together correctly? Correctness hinges on whether the heterogeneous machine satisfies the compound consistency model, i.e., memory operations from each device must satisfy the memory ordering rules of the device's consistency model.

We first discuss *hierarchical coherence*, wherein the two intra-device coherence protocols are stitched together via a higher-level inter-device coherence protocol. We then discuss how coarse-grained coherence tracking can help in mitigating the bandwidth demands of hierarchical coherence when used for multi-chip CPU-GPU systems. Finally, we will conclude with a simple approach to CPU-GPU coherence wherein blocks cached by the CPU are not cached in the GPU.

A Hierarchical Approach to Heterogeneous Coherence

Recall that in Section 9.1.6, we introduced hierarchical coherence for integrating two separate yet homogeneous coherence protocols. The same idea also can be extended for supporting heterogeneous coherence.

More specifically, hierarchical coherence works as follows. A local coherence controller, upon receiving a coherence request, attempts to fulfill the request locally within the device. Requests that cannot be completely fulfilled locally—e.g., a GetM request with sharers present in the other device—are forwarded to the global coherence controller, which in turn forwards the request to the other device's local coherence controller. Upon serving the forwarded request, that local controller responds back to the global coherence controller, which in turn forwards the response to the requestor.

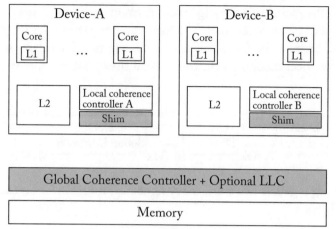

Figure 10.3: Integrating two or more devices into one heterogeneous machine via hierarchical coherence involves composing the two local coherence protocols together via a global coherence protocol.

To accomplish all of this:

- **the global controller** must be designed with an interface that is rich enough to handle coherence requests initiated by the devices' local controllers; and

- each of the local controllers must be extended with **shims** that not only serve as translators between the local and the global coherence controller interfaces, but also choose the appropriate global coherence requests to make, and interpret global coherence responses appropriately.

In order to better understand what the global coherence interface must look like, and the task of the shims, let us consider the following scenarios.

Scenario1: consistency-agnostic + consistency-agnostic.
Suppose we want to connect a multicore CPU with a CPU-like accelerator. Let us assume that both devices employ consistency-agnostic, SWMR-enforcing coherence protocols; however, the actual protocols employed by the two devices are different: the CPU uses a directory protocol whereas the accelerator uses snooping on a bus. Further, let us assume that the two protocols are stitched together via a global directory protocol with a coarse-grained sharer list, maintaining whether or not a cache block is present in one or more of the devices.

Now, suppose a CPU core performs a store to a block that is globally shared across the CPU and the accelerator. As shown in Fig. 10.4, the CPU core ① would send an Upg(rade) request to the local directory which must not only invalidate sharers within the CPU, but also ② send an Upg request to the global directory for invalidating sharers in the accelerator. Upon receiving the Upg, the global directory ③ must forward an Inv(alidation) request to the local snooping controller of the accelerator. The local snooping controller must interpret the Inv request as a GetM, because the GetM request is the one that invalidates sharers in a snooping protocol. ④ The local snooping controller would then issue the GetM on its local bus to invalidate sharers.

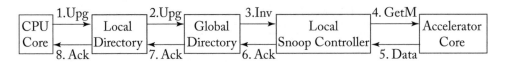

Figure 10.4: Stitching together two consistency-agnostic protocols: Coherence transactions for a store from a CPU core.

Thus, for stitching together consistency-agnostic protocols, the global coherence controller must have an interface that can handle requests and responses similar to Table 6.4 from Chapter 6. Further, the shims must issue requests to the global controller as per its interface, and interpret forwarded requests and responses from the global controller in accordance with the local controller's interface. For example, the snooping controller must interpret an Inv request from the global directory as a GetM.

Scenario2: consistency-directed + consistency-directed

Suppose we want to connect a GPU with another GPU-like accelerator via a shared LLC that serves as the global coherence controller. Let us assume that both the GPU and the accelerator enforce non-scoped release consistency using variants of release consistency-directed coherence protocols: the GPU employs the LRCC protocol whereas the accelerator employs the RCC protocol.

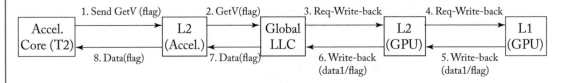

Figure 10.5: Stitching together two consistency-directed protocols.

Consider the message passing example shown in Table 10.10, assuming that T1 is on the GPU and T2 is on the accelerator. Let us assume that St1 and St2 have already been performed and the cache lines containing flag and data are in one of the GPU's L1s, with flag in O(wned) state. We show in Fig. 10.5 the sequence of coherence transactions that must be generated when T2 performs the acquire Ld1 in the accelerator. ① The accelerator would send a GetV request for flag to its local L2 controller. Since the accelerator's local L2 does not have the block containing flag, ② its shim must forward the GetV to the global LLC controller. The global LLC controller, upon finding flag to be owned by the GPU, ③ must send a request to the GPU's local L2 controller for writing back all dirty blocks. The GPU's L2 controller would ④ forward this request to the GPU L1 that owns the block. ⑤ The GPU L1 controller would then write back all of the dirty blocks (including flag and data) to the GPU's L2. ⑥ The GPU's L2, in turn, must write back the dirty blocks to the LLC, which is now the point of coherence. Then, ⑦ the global LLC controller must respond to the accelerator's L2 controller with the block containing flag. ⑧ Finally, the accelerator's L2 would respond to the requester, thereby completing the request.

Thus, with regard to stitching together consistency-directed protocols, the global LLC controller serves as the point of coherence. Therefore, the shims must forward requests (e.g., write-backs) to the LLC.

Scenario3: consistency-agnostic + consistency-directed.

Suppose we want to connect a CPU and a GPU together. Let us assume that the CPU enforces SC using a consistency-agnostic MSI directory protocol; the GPU enforces non-scoped RC using a consistency-directed LRCC protocol. Further, let us assume we want to connect the two via a global directory that is embedded in a globally shared LLC (Figure 10.3). How should the two coherence protocols be stitched together?

Consider the message passing example shown in Table 10.22, assuming that T1 is from the CPU and T2 is from the GPU. Initially, let us assume that both data and flag are cached in the L1s of the GPU as well as the CPU with initial values of 0.

Table 10.22: T1 from CPU adheres to SC. T2 from GPU adheres to non-scoped RC.

Thread T1 (CPU)	Thread T2 (GPU)	Comments
St1: St data1 = NEW; St2: St flag = SET;	Ld1: **Acq** Ld r1 = flag; B1: if(r1 ≠ SET) goto Ld1; Ld2: Ld r2 = data1	// Initially all variables are 0. Can r2==0?

When St1 is performed by the CPU, as shown in Figure 10.6, it would ① send an Upg request for data1 to the local directory controller which in turn ② must forward the request to the global directory. Importantly, the global directory need not forward the Upg to the GPU despite the GPU caching data1. This is because the LRCC protocol on the GPU does not require sharers to be invalidated upon a write since sharers are self-invalidated upon an acquire. Therefore, the global directory ③ can simply respond with an Ack. When the release (St2) is performed in the CPU, it must result in a similar sequence of coherence transactions.

When the acquire Ld1 from T2 is performed by the GPU, as shown in Fig. 10.7, it would ① send a GetV for flag to the GPU L2 controller which in turn ② must forward the request to the global LLC/directory. The global directory, upon finding that the block containing flag is

Figure 10.6: Stitching together a consistency-agnostic protocol with a consistency-directed one: Coherence transactions for St1 from the CPU.

in M(odified) state in the CPU, ③ must send a Fwd-GetS request to the CPU's directory. The CPU's directory must ④ forward that request to the CPU L1 that contains the block in state M. The CPU L1 would then ⑤ respond with the block containing flag, which must eventually be forwarded to the requestor.

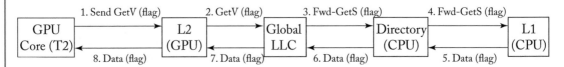

Figure 10.7: Stitching together a consistency-agnostic protocol with a consistency-directed one: Coherence transactions for Acquire Ld1 from the GPU.

Let us consider the same message passing example, assuming now that T1 is on the GPU and T2 is on the CPU, as shown in Table 10.23. When St1 is performed by the GPU, it simply writes data1 to its local L1.

Table 10.23: T1 from GPU adheres to non-scoped RC. T2 from CPU adheres to SC.

Thread T1 (GPU)	Thread T2 (CPU)	Comments
St1: St data1 = NEW;	Ld1: Ld r1 = flag;	// Initially all variables are 0.
St2: **Rel** St flag = SET;	B1: if(r1 ≠ SET) goto Ld1;	
	Ld2: Ld r2 = data1	Can r2==0?

When the release St2 is performed by the GPU, as shown in Fig. 10.8, the GPU would ① issue a GetO request for flag to the GPU L2 controller which in turn ② must forward it to the global directory. Because flag is cached in the CPU, the global directory ③ must forward an Inv request to the local directory controller of the CPU which would ⑥ respond after invalidating the cached copies of flag.

When Ld1 is performed by the CPU, as shown in Figure 10.9, ① a GetS request for flag would be sent to the local directory controller which in turn ② must forward it to the global LLC/directory controller. The global directory controller, upon finding that the block containing flag is owned by the GPU, ③ must send a request to the GPU's local L2 controller for writing

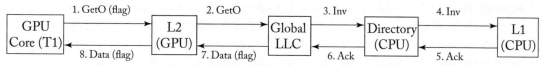

Figure 10.8: Stitching together a consistency-directed protocol with a consistency-agnostic one: Coherence transactions for St1 from the GPU.

back all dirty blocks. The GPU L2 controller must, in turn, ④ request a write-back of all dirty blocks from that L1. The GPU L1 controller, upon receiving the request, would ⑤ perform a write-back of its dirty blocks (including data1 and flag) to the GPU L2 controller. The GPU L2 controller must in turn ⑥ write back these blocks to the global directory/LLC. When the write-back for data1 reaches the global directory/LLC, the LLC upon finding that the block has sharers in the CPU, must ⑦ forward an Inv request for data1 to the CPU local directory, which in turn would ⑧ invalidate the L1 cached copy of data1 and respond with an ack. When the write-back for flag reaches the global directory/LLC, it must ⑨ forward the value to the L2 of the CPU, which in turn would ⑩ forward it to the L1, thereby completing the Ld1 request. Since data1 is now invalid in the L1 of the CPU, when Ld2 is performed by the CPU, it would get the correct value of NEW from the LLC.

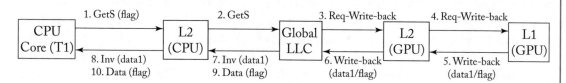

Figure 10.9: Stitching together a consistency-directed protocol with a consistency-agnostic one: Coherence transactions for Acquire Ld1 from the CPU.

The examples we considered in Scenario3 shed light on how the global coherence interface should behave. Any coherence request due to a GPU store must invalidate CPU sharers of that block. Therefore, for such requests, the global coherence interface must forward invalidations to the CPU, if there are any sharers. In contrast, any coherence request due to a CPU store need not invalidate GPU sharers since the GPUs are responsible for self-invalidating their cache blocks upon an acquire. In other words, the global coherence interface must disambiguate CPU sharers from GPU sharers. One way to realize this is to have two types of Read requests as part of the interface: (1) GetS, that requests a block for reading and asks the directory to track the block, so that the block can be invalidated upon a remote write; and (2) GetV [10], that requests a block for reading and is responsible for self-invalidating the block, and so the directory need not track the block.

Summary. Hierarchical coherence is an elegant solution to integrating heterogeneous coherence protocols. Hierarchical coherence requires: (1) a global coherence protocol with an interface that is rich enough to handle a range of coherence protocols; and (2) extensions to the original coherence protocols (shims) for interfacing with the global coherence protocol.

Is there a universal coherence interface that can be used to stitch together any two coherence protocols? Whereas the interface specified in Table 6.4 from Chapter 6 can handle any type of consistency-agnostic protocol, it falls short of handling a consistency-directed coherence protocol. Specifically, consistency-directed coherence protocols are characterized by read requests where the onus of invalidating the line lies with the requestor itself. A GetV, as opposed to a GetS, is a good match for such a read request. It is worth noting that a number of such coherence interfaces have been proposed both in academia (Crossing guard [22] and Spandex [10]) and industry (CCIX [2], OpenCAPI [4], Gen-Z [3], and AMBA CHI [1]).

Mitigating Bandwidth Demands of Heterogeneous Coherence

One problem with hierarchical coherence—especially for CPU-GPU multi-chip systems—is that the inter-chip coherent interconnect can become a bottleneck.

As the name suggests, in multi-chip systems the CPU and GPU are in separate chips and hence the global directory has to be accessed by the GPU via an off-chip coherent interconnect. Because GPU workloads typically incur high cache miss ratios, the global directory needs to be accessed frequently, which explains why the coherent interconnect can become saturated.

One approach [24] to mitigating this bandwidth demand is to employ coarse-grained coherence tracking [12], wherein coherence state is tracked at a coarse granularity (e.g., at page size granularity) locally in the GPU (as well as the CPU). When the GPU incurs a miss and the page corresponding to the location is known to be private to the GPU or read-only, there is no need to access the global directory. Instead the block may be directly accessed from memory via a high-bandwidth bus.

A Low-Complexity Solution to Heterogeneous CPU-GPU Coherence

CPUs and GPUs are not always designed by the same vendor. Unfortunately, hierarchical coherence typically requires modest extensions (shims) to both the CPU as well as the GPU protocol.

One simple approach to CPU-GPU coherence in multi-chip systems is *selective GPU caching* [7]. Any data that is mapped to CPU memory is not cached in the GPU. Furthermore, any data from the GPU memory that is currently cached in the CPU is also not cached in the GPU. This simple policy trivially enforces coherence. To enforce this policy, the GPU maintains a coarse-grained *remote directory* that maintains the data that is currently cached by the CPU. Whenever a block from the GPU memory is accessed by the CPU, that coarse-grained region is inserted into the remote directory. (If the GPU was caching the line, that line is flushed.) Any location present in the remote directory is not cached in the GPU.

Unfortunately, the above naive scheme can incur a significant penalty because any location that is cached in the CPU must be retrieved from the CPU. To offset the cost, several optimizations including GPU request coalescing (multiple requests to the CPU are coalesced), and CPU-side caching (a special CPU-side cache for GPU remote requests) have been proposed [7].

10.3 FURTHER READING

In this chapter we saw how consistency-directed coherence protocols are used for keeping GPUs coherent. Although most of the literature on CPU coherence uses the consistency-agnostic definition, there have been some classic works that have targeted coherence protocols to the consistency model.

Afek et al. [6] proposed lazy coherence that directly enforces SC without satisfying SWMR. Lebeck and Wood proposed dynamic self invalidation [18], the first self-invalidation based coherence protocol that targets both SC as well as weaker models directly without satisfying SWMR. Kontothanassis [17] proposed lazy release consistency for CPU processors, the precursor to similar such protocols proposed for the GPU.

The advent of multicores sparked renewed interest in consistency-directed coherence protocols. DeNovo [13] showed that targeting coherence toward DRF models can lead to simpler and scalable coherence protocols. VIPS [26] proposed a directory-less approach for directly enforcing release consistency, relying on TLBs for tracking private and read-only data. TSO-CC [14] and Tardis [31] are consistency-directed coherence protocols that directly target TSO and SC respectively.

A number of proposals for GPU coherence were adaptations of CPU coherence protocols. Temporal coherence [30] was the first to propose coherence for GPUs by adapting library coherence [19], a timestamp-based protocol for CPU coherence. HRF [15] extended DRF to a heterogeneous context and showed how scoped consistency models can be enforced. Meanwhile, Sinclair et al. [16, 29] adapted DeNovo for GPUs—specifically, by obtaining ownership for stores—and showed that non-scoped memory models can perform almost as well as scoped consistency models [16]. Alsop et al. [9] improved upon this by adapting lazy release consistency for GPUs. Finally, Ren and Lis [25] adapted Tardis for GPUs and showed that SC can be enforced efficiently on GPUs.

One challenge that we have not explicitly addressed is that accelerators, especially GPUs, often have programmer-controlled memories called *scratchpads* in addition to caches. There has been a body of work that has looked at integrating scratchpads as well as caches into the global address space [11, 16, 27].

10.4 REFERENCES

[1] The AMBA CHI Specification. `https://developer.arm.com/architectures/`

`system-architectures/amba/amba-5` 246

[2] The CCIX Consortium. `https://www.ccixconsortium.com/` 246

[3] The GenZ Consortium. `https://genzconsortium.org/` 246

[4] The OpenCAPI Consortium. `https://opencapi.org/` 246

[5] T. M. Aamodt, W. W. L. Fung, and T. G. Rogers. *General-Purpose Graphics Processor Architectures*. Synthesis Lectures on Computer Architecture. Morgan & Claypool Publishers, 2018. DOI: 10.2200/s00848ed1v01y201804cac044 212

[6] Y. Afek, G. M. Brown, and M. Merritt. Lazy caching. *ACM Transactions on Programming Languages and Systems*, 15(1):182–205, 1993. DOI: 10.1145/151646.151651 247

[7] N. Agarwal, D. W. Nellans, E. Ebrahimi, T. F. Wenisch, J. Danskin, and S. W. Keckler. Selective GPU caches to eliminate CPU-GPU HW cache coherence. In *IEEE International Symposium on High Performance Computer Architecture (HPCA)*, Barcelona, Spain, March 12–16, 2016. DOI: 10.1109/hpca.2016.7446089 246, 247

[8] J. Alglave, M. Batty, A. F. Donaldson, G. Gopalakrishnan, J. Ketema, D. Poetzl, T. Sorensen, and J. Wickerson. GPU concurrency: Weak behaviours and programming assumptions. In *Proc. of the 20th International Conference on Architectural Support for Programming Languages and Operating Systems (ASPLOS)*, pages 577–591, Istanbul, Turkey, March 14–18, 2015. DOI: 10.1145/2775054.2694391 214

[9] J. Alsop, M. S. Orr, B. M. Beckmann, and D. A. Wood. Lazy release consistency for GPUs. In *49th Annual IEEE/ACM International Symposium on Microarchitecture (MICRO)*, pages 26:1–26:13, Taipei, Taiwan, October 15–19, 2016. DOI: 10.1109/micro.2016.7783729 247

[10] J. Alsop, M. D. Sinclair, and S. V. Adve. Spandex: A flexible interface for efficient heterogeneous coherence. In *ISCA*, 2018. DOI: 10.1109/isca.2018.00031 245, 246

[11] L. Alvarez, L. Vilanova, M. Moretó, M. Casas, M. González, X. Martorell, N. Navarro, E. Ayguadé, and M. Valero. Coherence protocol for transparent management of scratchpad memories in shared memory manycore architectures. In *Proc. of the 42nd Annual International Symposium on Computer Architecture*, pages 720–732, Portland, OR, June 13–17, 2015. DOI: 10.1145/2749469.2750411 247

[12] J. F. Cantin, J. E. Smith, M. H. Lipasti, A. Moshovos, and B. Falsafi. Coarse-grain coherence tracking: RegionScout and region coherence arrays. *IEEE Micro*, 26(1):70–79, 2006. DOI: 10.1109/mm.2006.8 246

[13] B. Choi, R. Komuravelli, H. Sung, R. Smolinski, N. Honarmand, S. V. Adve, V. S. Adve, N. P. Carter, and C. Chou.   DeNovo: Rethinking the memory hierarchy for disciplined parallelism.   In *International Conference on Parallel Architectures and Compilation Techniques (PACT)*, pages 155–166, Galveston, TX, October 10–14, 2011. DOI: 10.1109/pact.2011.21 231, 247

[14] M. Elver and V. Nagarajan. TSO-CC: Consistency directed cache coherence for TSO. In *20th IEEE International Symposium on High Performance Computer Architecture (HPCA)*, pages 165–176, Orlando, FL, February 15–19, 2014. DOI: 10.1109/hpca.2014.6835927 247

[15] D. R. Hower, B. A. Hechtman, B. M. Beckmann, B. R. Gaster, M. D. Hill, S. K. Reinhardt, and D. A. Wood. Heterogeneous-race-free memory models. In *Architectural Support for Programming Languages and Operating Systems (ASPLOS)*, pages 427–440, Salt Lake City, UT, March 1–5, 2014. DOI: 10.1145/2541940.2541981 213, 217, 239, 247

[16] R. Komuravelli, M. D. Sinclair, J. Alsop, M. Huzaifa, M. Kotsifakou, P. Srivastava, S. V. Adve, and V. S. Adve. Stash: Have your scratchpad and cache it too. In *Proc. of the 42nd Annual International Symposium on Computer Architecture*, pages 707–719, Portland, OR, June 13–17, 2015. DOI: 10.1145/2872887.2750374 217, 231, 247

[17] L. I. Kontothanassis, M. L. Scott, and R. Bianchini.   Lazy release consistency for hardware-coherent multiprocessors.   In *Proc. Supercomputing*, page 61, San Diego, CA, December 4–8, 1995. DOI: 10.21236/ada290062 247

[18] A. R. Lebeck and D. A. Wood. Dynamic self-invalidation: Reducing coherence overhead in shared-memory multiprocessors. In *Proc. of the 22nd Annual International Symposium on Computer Architecture (ISCA)*, pages 48–59, Santa Margherita Ligure, Italy, June 22–24, 1995. DOI: 10.1109/isca.1995.524548 217, 247

[19] M. Lis, K. S. Shim, M. H. Cho, and S. Devadas. Memory coherence in the age of multicores. In *IEEE 29th International Conference on Computer Design (ICCD)*, Amherst, MA, October 9–12, 2011. DOI: 10.1109/iccd.2011.6081367 247

[20] D. Lustig, S. Sahasrabuddhe, and O. Giroux. A formal analysis of the NVIDIA PTX memory consistency model. In *Proc. of the 24th International Conference on Architectural Support for Programming Languages and Operating Systems (ASPLOS)*, 2019. DOI: 10.1145/3297858.3304043 228

[21] D. Lustig, C. Trippel, M. Pellauer, and M. Martonosi. ArMOR: Defending against memory consistency model mismatches in heterogeneous architectures. In *Proc. of the 42nd Annual International Symposium on Computer Architecture*, pages 388–400, Portland, OR, June 13–17, 2015. 240
DOI: 10.1145/2749469.2750378

[22] L. E. Olson, M. D. Hill, and D. A. Wood. Crossing guard: Mediating host-accelerator coherence interactions. In *Proc. of the 22nd International Conference on Architectural Support for Programming Languages and Operating Systems (ASPLOS)*, pages 163–176, Xi'an, China, April 8–12, 2017. DOI: 10.1145/3093336.3037715 246

[23] M. S. Orr, S. Che, A. Yilmazer, B. M. Beckmann, M. D. Hill, and D. A. Wood. Synchronization Using Remote-Scope Promotion. In *Proc. of the 22nd International Conference on Architectural Support for Programming Languages and Operating Systems (ASPLOS)*, pages 73–86, Istanbul, Turkey, March 14–18, 2015. DOI: 10.1145/2694344.2694350 217

[24] J. Power, A. Basu, J. Gu, S. Puthoor, B. M. Beckmann, M. D. Hill, S. K. Reinhardt, and D. A. Wood. Heterogeneous system coherence for integrated CPU-GPU systems. In *The 46th Annual IEEE/ACM International Symposium on Microarchitecture (MICRO-46)*, pages 457–467, Davis, CA, December 7–11, 2013. DOI: 10.1145/2540708.2540747 246

[25] X. Ren and M. Lis. Efficient sequential consistency in GPUs via relativistic cache coherence. In *IEEE International Symposium on High Performance Computer Architecture (HPCA)*, pages 625–636, Austin, TX, February 4–8, 2017. DOI: 10.1109/hpca.2017.40 226, 247

[26] A. Ros and S. Kaxiras. Complexity-effective multicore coherence. In *International Conference on Parallel Architectures and Compilation Techniques (PACT)*, pages 241–252, Minneapolis, MN, September 19–23, 2012. DOI: 10.1145/2370816.2370853 247

[27] Y. S. Shao, S. L. Xi, V. Srinivasan, G. Wei, and D. M. Brooks. Co-designing accelerators and SOC interfaces using gem5-aladdin. In *49th Annual IEEE/ACM International Symposium on Microarchitecture (MICRO)*, pages 48:1–48:12, Taipei, Taiwan, October 15–19, 2016. DOI: 10.1109/micro.2016.7783751 247

[28] D. E. Shasha and M. Snir. Efficient and correct execution of parallel programs that share memory. *ACM Transactions on Programming Languages and Systems*, 10(2):282–312, 1988. DOI: 10.1145/42190.42277 228

[29] M. D. Sinclair, J. Alsop, and S. V. Adve. Chasing away RAts: Semantics and evaluation for relaxed atomics on heterogeneous systems. In *Proc. of the 44th Annual International Symposium on Computer Architecture (ISCA)*, pages 161–174, New York, NY, ACM, 2017. DOI: 10.1145/3079856.3080206 247

[30] I. Singh, A. Shriraman, W. W. L. Fung, M. O'Connor, and T. M. Aamodt. Cache coherence for GPU architectures. In *19th IEEE International Symposium on High Performance Computer Architecture (HPCA)*, pages 578–590, Shenzhen, China, February 23–27, 2013. DOI: 10.1109/mm.2014.4 216, 217, 220, 223, 226, 247

[31] X. Yu and S. Devadas. Tardis: Time traveling coherence algorithm for distributed shared memory. In *International Conference on Parallel Architecture and Compilation (PACT)*, pages 227–240, San Francisco, CA, October 18–21, 2015. DOI: 10.1109/pact.2015.12 247

CHAPTER 11

Specifying and Validating Memory Consistency Models and Cache Coherence

By now we have hopefully convinced you that consistency models and cache coherence protocols are complex and subtle. In this chapter we discuss methods for rigorously specifying consistency models and coherence protocols and exploring their allowed behaviors. We also discuss methods for specifying their implementations and validating their correctness.

We start by briefly overviewing two methods for specifying concurrent systems—operational and axiomatic—focusing on how these methods can be used to specify consistency models and coherence protocols (Section 11.1). Given a formal consistency specification, we discuss how its behaviors can be explored automatically (Section 11.2). Finally, we will provide a whistle-stop tour of validating implementations, covering both formal and testing-based methods (Section 11.3)

11.1 SPECIFICATION

A specification serves as a contract between the user of the system and its implementer by defining precisely the set of legal behaviors of the system. A specification answers the question: What are the allowed behaviors of the system? In contrast, an implementation answers the question: How does the system enforce the prescribed behaviors?

What do we mean by behavior? A system interacts with its user (or environment) through a set of actions: (1) *input actions*: actions from the user to the system; (2) *internal actions*: those that happen within the system; and (3) *output actions*: actions from the system to the user. Among these only the input and the output actions are observable to the user—the internal actions are not. Thus, a sequence of (observable) input and output actions defines the behavior of the system. Of interest are two classes of behavioral properties: *safety* and *liveness*.

A safety property asserts that bad things should not happen—i.e., it specifies what sequences of observable actions are legal. For concurrent systems there are usually multiple legal sequences because of non-determinism. Safety is not enough to fully specify a system, however. Consider a system which accepts an input action and simply halts; although it has not exhibited anything bad, such a system is clearly not useful. That is why specifications also include liveness, which asserts that good things must eventually happen.

Before we explain how behaviors can be formally specified, let us consider what the observable actions are for consistency models and coherence protocols.

Consistency models: Observable actions and behavior

A memory consistency model is a contract between the software (user) and the hardware (implementer). Given a multithreaded program containing per-thread sequences of loads and stores, the memory model specifies what value each load must return. Thus, the relevant input actions are loads and stores (and their associated parameters including the processor core ID, address, and value to be stored in case of a store). The output actions are the values returned in response to the loads. A sequence of these observable actions (stores, loads, and the values returned) represents the behavior of a memory consistency model.

Coherence protocols: Observable actions and behavior

Recall that the processor core pipeline interacts with the coherence protocol to jointly enforce the desired memory consistency model. Thus, the "user" of the coherence protocol is the pipeline that interacts with it via two input actions: (1) read-request for every load in the program; and (2) write-request for every store in the program. The output actions from the coherence protocol are: (1) read-return that returns a value for every read-request; and (2) write-return that is simply an acknowledgment for the write. (The pipeline must know when a write completes.) A sequence of these observable actions represents the externally observable behavior of a coherence protocol.

It is important to note here the difference between what is observable in a coherence protocol vs. a consistency model. In the former, the instant when a read-request or write-request returns is observable. However, for a consistency model, the instant when a load or store returns is not observable—only the values returned for loads are observable.

Next, we discuss two major ways to specify the behavior of a system: *operational* and *axiomatic*. In the former, the system is described using an abstract reference implementation, whereas in the latter, mathematical axioms are used to describe the system's behaviors.

11.1.1 OPERATIONAL SPECIFICATION

An operational specification describes the behavior of a system using a reference implementation, typically expressed in the form of a state machine. The behaviors—i.e., the sequence of input/output actions—exhibited by the reference implementation *define* the set of all legal behaviors of the system. Operational models typically use internal states and internal actions to constrain the behavior of the system, thereby ensuring safety. Liveness in operational models hinges on the fact that state changes must eventually happen. Typically, liveness is expressed externally (outside of the state machine specification) using mathematical axioms written in *temporal logic* [31].

Specifying consistency models and coherence protocols operationally

Consistency models can be specified operationally using abstract implementations. Like their realistic counterparts, the abstract implementations typically have both the processor core pipeline component and the memory system component. Recall that the memory system component, which has the same interface as that of the coherence protocol (i.e., the same set of observable actions: read-request, read-return, write-request, write-return), specifies the coherence protocol.

In the following, we will describe two operational models for sequential consistency (SC). Both models have identical pipeline components but differ in their memory system components—the former specifies a consistency-agnostic coherence protocol whereas the latter specifies a consistency-directed coherence protocol. This will allow us to operationally characterize the differences between the two.

SC Operational Spec1: In-order pipeline + atomic memory

An example of an operational SC specification is similar to the naive SC implementation described in Chapter 3.6 (the switch). In addition to the observable actions—stores, loads and the values returned—the model employs internal state (memory) to constrain the values that can be read by the loads.

The operational specification works as follows. (It is assumed that each core has a program counter, pointing to the next instruction to be fetched.)

Step 1: *Fetch*: One of the processor cores is non-deterministically selected; it fetches its next instruction and inserts it into a local instruction queue.

Step 2: *Issue from the pipeline.* Again one core is non-deterministically selected; it decodes the next instruction from its instruction queue. If it is a memory instruction, the pipeline issues a read-request (for a load) or a write-request (for a store) and blocks.

Step 3: *Atomic memory system.* Upon receiving a read-request, it reads the location from memory and responds with the value; upon receiving a write-request, the memory system writes it to memory and responds with an ack.

Step 4: *Return to the pipeline:* The pipeline unblocks upon receiving a response: ack (in case of a store) or value (in case of a load).

The behaviors produced by this specification are sequences of input actions in program order (load/store at Step 1) and output actions (value returned for a load at Step 4). It is easy to see that its behaviors satisfy SC. For example, in the message passing example in Table 11.1 the following sequence is observable:

«S1, S2, L1:r1=SET, L2:r2=NEW»

Likewise, several other sequences in which L1 repeats one or more time until it sees a value of SET are possible. But the following SC violation is not observable:

Table 11.1: Message passing example

Core C1	Core C2	Comments
S1: St data = NEW; S2: St flag = SET;	L1: Ld r1 = flag; B1: if(r1 ≠ SET) goto L1; L2: Ld r2 = data;	// Initially data and flag are 0. L1, B1 may repeat many times.

«S1, S2, L1:r1=SET, L2:r2=0»

SC Operational Spec2: In-order pipeline + buffered memory

Now let us consider another SC operational model in which the atomic memory is now front-ended by a global FIFO store queue to form a buffered memory system. Each entry in the store queue contains the address, the value to be stored, and the ID of the core that issued the write request. The buffered memory system responds to a write request with two types of acks: an ack to denote that the request has been inserted into the write queue and another late-ack when the write has been written to memory. More specifically, the buffered memory system works as follows. (Steps 1 and 4 are identical to the previous spec and hence omitted.)

Step 2': *Issue from the pipeline.* One core is non-deterministically selected; it decodes the next instruction from its instruction queue. If it is a store instruction, the pipeline issues a write-request and blocks; if it is a load instruction, the pipeline waits for the latest write to be late-acked, then issues a read-request and blocks.

Step 3': *Buffered memory system.* Upon receiving a write-request, the buffered memory system inserts it into the global store queue and responds to the pipeline with an ack. (When the write eventually gets written to memory, a late-ack is sent to the processor core that initiated the write.) Upon receiving a read-request, the value is read from memory and returned to the pipeline.

Despite the buffering, the behavior produced by the model satisfies SC. Returning to the example shown in Table 11.1:

«S1, S2, L1:r1=SET, L2:r2=0»

The above SC violation is not observable because the global store queue is a FIFO and hence commits stores to memory without violating per-core program order.

Consistency-agnostic vs. Consistency-directed coherence

Recall that in Chapter 2 we classified coherence interfaces into two categories: consistency-agnostic and consistency-directed. The atomic memory system (Step 3 from Spec1) specifies a consistency-agnostic protocol since writes happen synchronously—whereas the buffered memory system (Step 3') is one example of a consistency-directed protocol specification since writes are asynchronous. (More aggressive specifications of the latter are possible [3].) Although both of these specification combine with an in-order pipeline to enforce SC, the two protocols show different behaviors. Consider the sequence of externally observable actions shown in Table 11.2.

«write-request(X,1), write-return(X,1),read-request(X), read-return(X,0)»

Table 11.2: Linearizability vs. SC: whereas SC allows for the read to return a 0, linearizability does not.

Time	Core C1	Core C2
t_0	Write-request (X,1)	
t_1	Write-return (X,1)→ok	
t_2		Read-request (X)
t_3		Read-return (X)→0

The above sequence cannot be observed in an atomic memory system although the sequence does not violate SC. This is because the atomic memory specification ensures that the write is actually written to memory between its invocation (t_0) and response (t_1), ensuring that the read will return a 1 and not a 0. However, the buffered memory specification allows this behavior because the write to X can be buffered in the global queue when the read happens, allowing the read to see a 0.

The atomic memory system (or consistency-agnostic coherence) can be modeled using a correctness condition that is stronger than SC called linearizability [16]. In addition to the SC rules, linearizability requires that a write must take effect in real time between its invocation and response. Linearizability is a *composable* property: an object is linearizable iff its components are linearizable. Translating to our setting, a memory system is linearizable iff each of its locations are individually linearizable. Thus, consistency-agnostic coherence can be specified on a per-memory location basis, i.e., per-memory location linearizability. This also explains why coherence of individual locations can be enforced independently using, for example, distributed directories.

In contrast, SC is not a composable property. Even if all memory locations satisfy SC independently (and the pipeline does not reorder memory operations), it does not imply that the whole system is SC. Recall that in Chapter 2 (sidebar) we looked at consistency-like definitions of coherence, where coherence was defined as SC on a per-memory location basis. Such a definition of coherence, however, does not completely specify a coherence protocol. (Because linearizability is stronger than SC, per-location SC is not strong enough to specify consistency-agnostic coherence. Because SC is not compositional, per-location SC is not strong enough to specify consistency-directed coherence as well.) Nevertheless, per-memory location SC is a useful safety check for coherence. Any consistency-agnostic protocol—recall that it satisfies per-memory location linearizability—will necessarily satisfy per-memory location SC . Even for a consistency-directed coherence protocol, per-location SC is a useful safety check since most consistency models explicitly mandate SC on a per-location basis (and call it the "coherence" axiom! [4]).

In summary, whereas consistency-agnostic coherence exposes an atomic (linearizable) memory interface, consistency-directed coherence exposes an interface of the target consistency model.

Specifying implementations operationally

Until now, we saw how consistency models and coherence protocols can be specified operationally. One of the benefits of the operational method is that lower level implementations (e.g., detailed coherence protocols) can be modeled naturally by refining the base specification with implementation-specific internal states and actions.

For example, let us add caches to the SC operational specification (Spec1) described above, with each processor now holding a local cache and its associated coherence controller. Each read/write request from the pipeline is now sent first to the local cache controller that checks for a hit or a miss. If it is a miss, the cache controller contacts the directory/memory controller with a GetS or GetM request (as we saw in the earlier chapters on coherence protocols). The exact state transitions naturally depend on the specifics of the coherence protocol employed, but the takeaway here is that the coherence protocols discussed earlier in table format are essentially operational models obtained by refining the memory system (Step 3 or Step 3'). In a similar vein, it is possible to refine the pipeline specification to model additional pipeline stages to faithfully model the pipeline implementation.

Tool support

Operational models can be directly expressed in any language for expressing state machines such as Murphi (http://mclab.di.uniroma1.it/site/index.php/software/18-cmurphi) or TLA (https://lamport.azurewebsites.net/tla/tla.html). In particular, each of the coherence protocol tables in the book are easily expressed as state machines.

11.1.2 AXIOMATIC SPECIFICATION

Axiomatic specification is a more abstract way of specifying the behavior of a concurrent system, whereby mathematical axioms composed of observable actions (and relations between these actions) are used to constrain the set of allowed behaviors.

Specifying consistency models axiomatically

The formalism we used to describe the SC, TSO, and XC memory models in the earlier chapters corresponds to the axiomatic method. Recall that the formalism is composed of the following components.

- *Observable actions*: These are the loads, stores and the values returned for the loads.

- *Relations*: The program order relation is defined as the per-core total order representing the order in which loads and stores appear in each core. The global memory order is defined as a total order on the memory operations from all cores.

- *Axioms*: There are two categories of axioms—*safety* and *liveness*—that specify the safety and liveness properties, respectively. For SC there are three safety axioms: (1) preserved program order axiom—the global memory order respects program order of each core; (2) load value axiom—the load gets the value written by the most recent store before it in global memory order; and (3) the atomicity axiom—the load and store of an RMW instruction occur consecutively in global memory order. In addition to these safety axioms, there is also a liveness axiom that states that no memory operations must be preceded by an infinite sequence of other memory operations; informally, this means that a memory operation must eventually perform and cannot be delayed indefinitely.

For any sequence of observable actions—if we are able to construct a global memory order that adheres to the above axioms, then it said to be a legal behavior of the memory model.

Specifying coherence protocols axiomatically

We saw earlier how coherence protocols can be specified operationally. Here, we see how they can be specified at a more abstract level axiomatically, which can be useful for verification purposes. Specifically, we focus on specifying a consistency-agnostic coherence protocol that satisfies linearizability. (Consistency-directed coherence protocols can be specified similarly to how consistency models are specified axiomatically.) Recall that linearizability is stronger than SC, e.g., it disallows the sequence of actions shown in Table 11.2. Therefore, the specification adds an additional axiom over and above SC to constrain such behavior.

- *Observable actions*: Because we are specifying a coherence protocol, the relevant actions are the read-request, read-return, write-request, and write-return events (the coherence protocol interface as seen by the processor pipeline).

- *Internal action:* In addition to the observable actions, we add two internal actions—read-perform and write-perform—the instants in which a read or a write takes effect.

- *Relations*: Like SC, the global memory order is defined to be a total order on the read-perform and write-perform events from all cores.

- *Axioms*: In addition to the three safety axioms that are associated with SC there is a fourth axiom that says that a read or write must perform between its invocation and response; more formally, a write-perform (read-perform) action must appear in between the write-request (read-request) and write-return (read-return) actions in the global memory order. Finally, like in SC there is a liveness axiom that says that any read or write request must eventually return.

Specifying implementations axiomatically

We saw earlier how implementations (e.g., coherence protocol implementations) can be naturally expressed operationally by extending the base operational specification with internal states and actions. In a similar vein, implementations can also be expressed axiomatically by extending the base axiomatic specification. In this section, we focus on how a processor core pipeline can be specified axiomatically following Lustig et al. [20].

Recall the abstract nature of the axiomatic specification, where a load/store is modeled as a single instantaneous action. Our goal now is different though. It is not to specify correctness—rather, the goal is to faithfully model a realistic processor core in which each load or store goes through several pipeline stages. Therefore, a single load or store action is now expanded into multiple internal pipeline sub-actions with each sub-action representing a pipeline stage.

For example, let us consider the classic five-stage pipeline. Here, a load is split into five sub-actions: fetch, decode, execute, memory, and writeback. A store is split into seven actions: fetch, decode, execute, memory, writeback, exit-store-buffer and memory-write. ("Memory" refers to the memory stage in the pipeline, whereas "memory-write" refers to the sub-action in which the value is actually written to memory.)

Recall that a key component of the axiomatic consistency specification is the preserved program order (ppo) axiom, that mandates that program order must be preserved in the global memory order. Analogous to ppo, the idea is to use *pipeline ordering axioms* for specifying microarchitectural happens-before—i.e., specifying whether or not the pipeline preserves the ordering between sub-actions of different instructions. Again, it is important to understand the subtle difference between ppo and the pipeline ordering axioms. Whereas the goal of the former is to specify correctness, the goal with the latter is to faithfully model the pipeline. In fact, whether or not the pipeline implementation honors the ppo axioms (for the intended consistency model) is an important validation question, which we address in the next section.

For the five-stage pipeline the pipeline ordering axioms are as follows.

- *Fetch honors program order.* If instruction i_1 occurs before i_2 in program order (po), then the fetch of i_1 happens before i_2. That is: $i_1 \xrightarrow{po} i_2 \implies i_1.fetch \to i2.fetch$

- *Decode is in order.* If fetch of i_1 happens before i_2, then decode of i_1 happens before i_2. That is: $i_1.fetch \to i2.fetch \implies i_1.decode \to i2.decode$

- In a similar vein, execute, memory and writeback stages preserve ordering (elided).

- *FIFO store buffer.* If writeback of store i_1 happens before that of store i_2, then i_1 exits the store buffer before i_2. That is: $i_1.writeback \to i2.writeback \implies i_1.exits\text{-}store\text{-}buffer \to i2.exits\text{-}store\text{-}buffer$

- *Ordered writes.* If i_1 exits the store buffer before i_2, then i_1 writes to memory before i_2. That is: $i_1.exits\text{-}store\text{-}buffer \to i2.exits\text{-}store\text{-}buffer \implies i_1.memory\text{-}write \to i_2.memory\text{-}write$.

Finally, the load value axiom specifies the constraints under which a load can read the value from a store of the same address.

- *Load value axiom.* If a store i_1 writes to memory before load i_2's memory stage (and there are no other memorywrites in the middle from other stores), then i_2 reads the value written by i_1.

Any sequence of pipeline actions (and load values returned) that adhere to the above axioms are legal behaviors. In order to understand this better, consider the message passing example shown in Table 11.3. When L1 sees the new value SET, can L2 see the old value 0? The five-stage pipeline would permit this behavior if a global sequence of pipeline actions can be constructed that satisfies the axioms above. If not—i.e., if the constraints lead to a cycle, the behavior is not permitted. Let us attempt to construct a global sequence.

- $S1.memory\text{-}write \rightarrow S2.memory\text{-}write$ (the pipeline ordering axioms ensure that S1 writes to memory before S2).

- $S2.memory\text{-}write \rightarrow L1.memory$ (S2 has to be written to memory before L1's memory stage—only then, will L1 be able to read from S2).

- $L1.memory \rightarrow L2.memory$ (the pipeline ordering axioms ensure that L1 performs its memory stage before L2).

- $L2.memory \rightarrow S1.memory\text{-}write$ (if L2 must read the old value, and not the one produced by S1, its memory stage must have happened before S1's memory write).

Table 11.3: Message passing example

Core C1	Core C2	Comments
S1: St data = NEW;	L1: Ld r1 = flag;	L1 reads SET
S2: St flag = SET;	L2: Ld r2 = data;	Can L2 read initial value 0?

As we can see, the constraints listed above are impossible to satisfy as it results in a cycle. Thus, the five-stage pipeline does not permit this behavior for this example.

In the above discussion, we had not modeled caches or cache coherence explicitly. Manerkar et al. [24] show how the axiomatic specification of the pipeline can be extended to model the coherence protocol/pipeline interactions.

In summary, we saw how pipeline implementations can be specified axiomatically. The primary benefit of modeling implementations is so that they can be validated. In Section 11.3.1 we will see how axiomatic models of the implementation can be validated against its consistency specifications.

Tool support.

Specification languages such as *Alloy* (`http://alloy.mit.edu`) or *Cat*, associated with the *Herd* tool (`http://diy.inria.fr/herd/`), can be used to express axiomatic specifications of consistency models. The μspec domain specific language allows for pipeline and coherence protocol implementations to be specified axiomatically (`http://check.cs.princeton.edu/#tools`).

11.2 EXPLORING THE BEHAVIOR OF MEMORY CONSISTENCY MODELS

Because consistency models define shared memory behavior, it is crucial for programmers and middleware writers—e.g., compiler writers and kernel developers—to be able to understand what behaviors are permitted by the memory model and what behaviors are not. Typically, memory consistency models are specified by processor vendors in prose, which can often be ambiguous [1]. A formally specified memory consistency model is not only unambiguous, but also enables behaviors to be explored automatically. But before delving into how this exploration can be carried out, let us briefly discuss *litmus tests*—simple programs that reveal the properties of a memory model.

11.2.1 LITMUS TESTS

A memory consistency model is likely not very relevant to a high-level programmer who uses libraries for synchronization. It is, however, important for low-level programmers who develop synchronization constructs—be it a compiler writer who implements language level synchronization or a kernel writer who develops kernel level synchronization or an application programmer who develops lock-free data structures. Such low-level programmers will want to know the behaviors exhibited by the memory model so that they can get the desired behavior for their hand-crafted synchronization situation. Litmus tests are simple multithreaded code sequences that abstract synchronization situations; the behavior(s) exhibited by the memory model in these tests are thus instructive to the low-level programmer. For instance, they may reveal whether or not a particular FENCE is required to get the desired behavior. These litmus tests are also used in processor vendor manuals to explain the properties of the memory consistency model.

Examples.

Recall that we used such litmus tests to explain the behavior of memory models in an earlier chapter. Table 3.3 for instance is dubbed Dekker's litmus test since it abstracts the synchronization situation in Dekker's algorithm for mutual exclusion; Table 3.1 is often referred to as the message passing test. Consider another example shown in Table 11.4 which is referred to as the "coherence" litmus test. An execution that results in r1 reading the NEW value but r2 reading the old value of 0 violates the data-value invariant and hence violates the coherence litmus test.

Table 11.4: Can r1 be NEW while r2 is 0?

Core C1	Core C2	Comments
S1: x = NEW;	L1: r1 = x;	// Initially x = 0
	L2: r2 = x	

11.2.2 EXPLORATION

Given a formal specification of a memory model and a litmus test, can we automatically explore all possible behaviors of the memory model for that litmus test? Fortunately, there are tools that do this. Intuitively, once a memory model is formalized—axiomatically or operationally—it is possible to exhaustively enumerate every possible schedule of the litmus test and, for each such execution, employ the mathematical axioms (in case of axiomatic) or execute the state machine (in case of operational) to determine the output of the execution (the values returned by loads).

Tool support

Recall that the *herd* tool provides a language to express consistency models mathematically; the tool also has a built-in simulator for exploring behaviors of litmus tests. The ppcmem tool (`https://www.cl.cam.ac.uk/~pes20/ppcmem/help.html`) does something similar but is specialized for POWER and ARM memory models, while the CppMem tool (`http://svr-pes20-cppmem.cl.cam.ac.uk/cppmem/`) is geared toward the C/C++ language-level memory model. As far as operational models are concerned, the RMEM tool (`https://www.cl.cam.ac.uk/~sf502/regressions/rmem/`) has built-in operational models for ARM (multiple variants), POWER (multiple variants), x86-TSO and RISC-V, and also allows for their behaviors to be explored.

How to generate litmus tests? They can of course be generated manually to test any interesting memory model feature or synchronization situation. There is also tool support. Litmus tests can be generated randomly via a test generator [14, 15]. The diy tool (`http://diy.inria.fr/doc/gen.html`) helps generate a litmus test given the shape of the test (the program order relations and the shared memory dependencies). Finally, litmusttestgen (`https://github.com/nvlabs/litmustestgen`) is a more automated method to generate a comprehensive set of litmus tests for a memory model specification expressed axiomatically.

While litmus tests are a good way to convey intuition about a memory model, they cannot normally serve as a complete specification of the memory model, since they leave some potential behaviors (those that are not covered in the tests) unspecified. But given a sufficient number of tests, and a syntactic template of the memory model (with missing pieces), MemSynth (`http://memsynth.uwplse.org/`) shows how a complete memory model satisfying the litmus tests can be synthesized.

11.3 VALIDATING IMPLEMENTATIONS

Memory consistency enforcement spans multiple sub-systems including the processor pipeline and the cache and memory sub-systems. All of these sub-systems must cooperate with each other to enforce the promised memory consistency model. Validating that they do so correctly is the topic of this section.

Exhaustively validating that the complete implementation satisfies the promised memory model is the end goal. But the sheer complexity of modern multiprocessors makes this a very hard problem. Therefore there is a gamut of validation techniques, ranging from formal verification to informal testing. This is a huge topic in itself and probably worthy of its own book. Here, we provide a flavor of some of these techniques with pointers for further study.

11.3.1 FORMAL METHODS

In this category, a formal model of an implementation is validated against a specification. We classify these methods based on whether the specification and the implementation are expressed axiomatically or operationally. We then explain each category by considering one or two examples of techniques belonging to that class. We conclude this section with some closing thoughts on operational vs. axiomatic methods.

Operational Implementation against Axiomatic Specification

Manual proof
Given an implementation in the form of an operational model and the specification in the form of an axiomatic model, one can show via a manual proof that the implementation satisfies each of the axioms of the specification.

For example, consider a system in which the processor pipeline presents loads and stores in program order and a cache coherence protocol that ensures SWMR and data-value invariants. Meixner and Sorin [26] showed that such a system enforces SC by proving that it satisfies each of the SC axioms.

One powerful template [30] for coming up with a proof follows Lamport's method for totally ordering the events of a distributed system [18]. The idea is to assign hypothetical timestamps for every independent action (e.g., a fetch of a store instruction at the pipeline) and derive timestamps of causally related events (e.g., a GetM request due to the store gets a higher timestamp). In doing so, a global order of all memory operations is derived. Then every value loaded is checked to determine whether it returns a value consistent with the load value axiom.

Model checking
Operational specifications of coherence protocols are commonly verified against axiomatic invariants using model checkers. Model checkers, such as Murphi [12], allow for coherence protocols similar to those described in tabular format in the earlier chapters to be expressed in a domain specific language. The language also allows for axioms such as SWMR to be expressed.

Then the model checker explores the entire state space of the protocol to either ensure that the invariants hold or report a counter example (a sequence of states that violates the invariant). However, such explicit-state model checking can only be carried out for limited models of the implementation (e.g., two or three addresses, values, and processors) because of state-space explosion.

There are a number of approaches for countering state-space explosion. Symbolic model checkers [7] use logic for manipulating sets of states (rather than single states), thereby enabling the scaling of the models. Bounded model checkers [6] compromise on exhaustiveness by only looking for counter examples of a limited length in the sequence of states. Finally, one approach for a complete proof is to model-check exhaustively for a limited system and then employ *parameterization* [11, 17] to generalize for an arbitrary number of processors, addresses etc.

The following is a concrete example of how a non-trivial coherence protocol can be model-checked using Murphi: (`https://github.com/icsa-caps/c3d-protocol`)

Operational Implementation against Operational Specification

Suppose there are two state machines: Q (the implementation) and S (the specification) with the same set of observable actions. How do we prove that Q implements S faithfully? In order to establish this, we need to prove that all of the behaviors—the sequences of observable actions—of Q are contained in S.

Refinement

One stylized proof technique that allows one to reason about pairs of states in Q and S (rather than sequences) is called refinement [2] (also known as simulation [27]). The key idea is to identify an abstraction function F (also known as refinement mapping or simulation relation) that maps every reachable state of the implementation to that of the specification, such that:

- initial states ($q_0 \in Q, s_0 \in S$) are part of the mapping. That is, $F(q_0) = s_0$; and

- the abstraction function preserves state transitions. That is, for every state transition in the implementation between two states q_i and q_j with some observable action a, i.e., $q_i \xrightarrow{a} q_j$, the abstraction function will lead to two states in the specification with the same observable action, i.e., $F(q_i) \xrightarrow{a} F(q_j)$.

For example, consider a cache coherent system containing caches and global memory kept coherent using an MSI coherence protocol on an atomic bus (the implementation). Recall that the overall state of the implementation includes the states of all the caches (value and the MSI state for every location) as well as the state of the global memory. We can show that the cache coherent memory implements atomic memory (the specification) using refinement. To this end, we identify an abstraction function relating the state of cache coherent memory to the state of atomic memory. Specifically, the value of an atomic memory block for a given location is given by: (1) the value of the cache block for that location, if the block is in M state; (2) otherwise, it is identical to the values in each of the blocks that are in S state and the global memory. The proof would then look at all possible state transitions in the protocol to check whether the abstraction function holds. For example, consider the $S \xrightarrow{Write(X,1)} M$ transition for some cache block X whose initial value is 0. For the cache block that goes into M state the abstraction function holds, as it will cache the up-to-date value of 1. For the other cache blocks that were in state S, the abstraction functions holds provided the MSI protocol correctly enforces SWMR and invalidates all those blocks.

Thus, a refinement proof not only requires the identification of an abstraction function, it also requires subsequent reasoning about the state transitions in the implemented system; that process might reveal other invariants (e.g., SMWR revealed above) that need to be proven. The Kami framework [10] uses a refinement proof to show that an operational specification of a processor combined with a cache coherent memory system satisfies SC.

Model checking

One way to show that an implementation satisfies the specification is to run the two state machines side by side with the same sequence of input actions, and observe whether they produce the same output actions. While this simple strategy is sound, it may not always succeed. (When the method says that the state machines are equivalent they are definitely equivalent—but when it says they are not equivalent, they may yet turn out to be equivalent.) This is because of the non-determinism inherent in concurrent systems—sometimes, the specification and the implementation may show identical behavior, but with different schedules. Having said this, this is quite a useful strategy, especially when the implementation and specification are conceptually similar. Banks et al. [5] employed this strategy to show that a cache coherence protocol satisfies the promised consistency model.

Axiomatic Implementation Against Axiomatic Specification

We saw how implementations (such as the core pipeline implementation) can be specified axiomatically. How to formally validate that such an axiomatic model satisfies the axiomatic specification[1] of the consistency model?

The Check line of work [20, 21, 24], leverages the idea of exhaustive exploration for this purpose. Given a litmus test and an axiomatic specification of the memory model, we saw in Section 11.2 how all of the behaviors of the memory model on the litmus test can be exhaustively explored. Such an exhaustive exploration can be carried out for axiomatic specifications of implementations as well. The behaviors on the litmus test can then be compared against the behaviors on the abstract axiomatic model. If the implementation shows more behavior, it indicates a bug in the implementation. On the other hand, if the implementation shows less behavior than the specification, it indicates that the implementation is conservative. The same process is then repeated for other litmus tests that are part of the suite.

---

[1]This section assumes validation against an axiomatic specification. A similar approach validating an axiomatic model of an implementation against operational specification is possible, although we know of no such proposal in this category.

One limitation of the above approach is that the validation is exhaustive only for the litmus tests explored. A complete verification requires exploring all possible programs. PipeProof [22] accomplishes this with an induction style proof on a symbolic abstraction of a program which allows for programs of any number of instructions and all possible addresses and values to be modeled.

Summary: Operational vs. Axiomatic

Generally speaking, axiomatic specifications are declarative; they describe what behaviors are permitted without fully describing how the system is implemented. Typically, they are more abstract and more easily expressed in mathematics; thus, it is faster to explore their behaviors. On the other side of the coin, operational models are closer to implementations and more intuitive for an architect; thus, it is arguably easier to model detailed implementations operationally. Recall that we were able to naturally describe fairly complex coherence protocols in the earlier chapter in table format. In summary, whether to go down the axiomatic or operational route is not a question that has a simple answer.

Finally, it is important to understand that hybrid axiomatic/operational models are also possible. For example, let us consider a multiprocessor with caches kept coherent using a coherence protocol. Instead of describing the coherence protocol operationally, it is also possible to abstract it axiomatically using the SWMR and the data-value invariants. Such an approach might be attractive when breaking down a complex verification problem into simpler ones—for example, when verifying that the combination of the pipeline and the memory system correctly enforces the consistency model, it might make sense to verify the coherence protocol independently and then abstract it axiomatically via SMWR.

11.3.2 TESTING

Formally proving that the implementation satisfies the specification provides high confidence that the implementation is indeed correct. Yet, formal verification is not a panacea. The reasons are threefold. First, formal verification is hard to scale. Automatic techniques such as model checking may not be able to handle the humongous state spaces of complex implementations; on the other hand, techniques that can handle such large state spaces are not fully automated and are not easy for architects. Second, formally specifying an implementation correctly is not straightforward and may be error prone. Third, even if the model is accurate, it may be incomplete, in the sense that there might be parts of the implementation that the model does not consider. That is why testing the final implementation thoroughly is extremely important. We classify testing into two categories: *offline testing* and *online testing*. In the former (also known as static testing), a real or simulated multiprocessor is tested for consistency violations prior to being deployed. With the latter, violations are detected at runtime after deployment.

Offline Testing

The idea is to run test programs on a real (or simulated) machine, observe their behaviors, and validate them. The approach raises two questions: How are the test programs generated? How are their behaviors validated?

One of the earlier works on memory consistency testing, TSOtool [15], uses a pseudo-random generator to generate short test programs. Given a previously unseen randomly generated program, how to validate its observed behavior (values returned by reads)? This is the key challenge that TSOtool addresses. Intuitively, this demands a *consistency checker* that dynamically reconstructs a global order from the observations and checks whether the order is allowed by the consistency model While reconstructing a global order from values returned by reads is an NP-complete problem, TSOtool comes up with a polynomial algorithm that trades off accuracy for performance (it might miss some violations). MTraceCheck [19] improves upon this checker further.

Instead of using randomly generated programs, one can use litmus tests (Section 11.2.1). With litmus tests, the advantage is that a suite of tests is already available for most memory models; moreover, for each litmus test the expected behavior is known, which obviates the need for a sophisticated checker that reconstructs the global memory order. The litmus tool diy (http://diy.inria.fr/doc/litmus.html) is a framework for testing real processors using this approach.

While litmus testing and random testing are effective in a post-silicon environment, they may not be that effective in a slow simulator environment. McVersi [14] proposes a genetic-programming-based approach that leverages the white-box nature of the simulator environment to come up with test programs that are likely to reveal consistency violations faster.

Online Testing

Offline testing is certainly beneficial but, because of the fundamental limitation of testing, it may fail to uncover some bugs. Furthermore, even if offline testing did not miss any bugs, hardware transient errors and fabrication errors may yet cause consistency violations in the wild. In online testing (also known as dynamic testing), hardware support is added to a multiprocessor for detecting such violations during execution. One conceptually straightforward way to do this is to implement a consistency checker as in TSOtool, but in hardware. However, a naive implementation of such a scheme would be impractically expensive in terms of memory (metadata) and communication overhead. Chen et al. [9] propose a scheme to reduce these costs significantly, leveraging the fact that not all memory operations need to be tracked (e.g., loads and stores to local variables need not be tracked). Meixner and Sorin [25] propose an alternative strategy that reduces the verification task into that of verifying two invariants: (a) checking that the cache coherence protocol correctly enforces SWMR; and (b) checking that the pipeline correctly interacts with the coherence protocol. Notably, they show that these two invariants can be efficiently checked in hardware.

11.4   HISTORY AND FURTHER READING

In the same paper that Lamport defined SC, he also provided (its first) operational specification. The specification is a variant of, and semantically equivalent to, our SC Operational Spec1 in Section 11.1.1. It is interesting to note that the very first operational specification of SC employed a linearizable memory system. (Herlihy and Wing [16] formalized linearizability a decade later.) The first operational specification of SC that used a consistency-directed (i.e., non-linearizable) coherence protocol was proposed by Afek et al. [3]. Shasha and Snir [34] were the first to provide an axiomatic specification of SC.

The 1990s and early 2000s saw significant research on formally verifying coherence protocols, not only against invariants such as SWMR but also against memory consistency models [32]. Park and Dill [29] showed how the protocol used in the FLASH multiprocessor protocol can be abstracted and verified using a theorem prover. Qadeer [33] showed that verifying against SC is undecidable, but can be made tractable for most practical protocols. Chatterjee et al. [8] showed how the pipeline combined with a coherence protocol can be automatically verified against an operational specification of a weak memory model in a model checker.

Because of the complexity of formally verifying coherence protocols—state space explosion combined with the complexity of parameterization—there have been efforts to design coherence protocols in a way that enables straightfoward parameterized verification for an indefinite number of processors [37, 38]. Another promising direction is a correct-by-construction approach to coherence protocols. TRANSIT [36] and verC3 [13] leverage program synthesis technology for auto-completing coherence protocols given a partial implementation. ProtoGen [28] proposes a method for automatically producing a complete and concurrent protocol (with transient states and actions) given an atomic protocol with only stable states and transitions.

In this chapter, we focused on methods for validating whether or not the microarchitecture specification (e.g., pipelines and coherence protocols) satisfies the architectural consistency specification. However, a complete validation requires spanning the stack from high-level languages, (through compilers, ISA, microarchitecture) to RTL. TriCheck [35] validates that the HLL, compiler, ISA, and microarchitecture satisfy the language-level consistency specification for a set of litmus tests. RTLCheck [23] validates that the RTL faithfully implements the microarchitectural consistency specification for a set of litmus tests.

To conclude, we must stress here that we have but scratched the surface on this topic that is, despite its long history, continuing to see exciting new results.

11.5   REFERENCES

[1] Linux Kernel mailing list: Spin unlock optimization. https://lists.gt.net/engine?post=105365;list=linux 262

[2] M. Abadi and L. Lamport. The existence of refinement mappings. *Theoretical Computer Science*, 82(2):253–284, 1991. DOI: 10.1109/lics.1988.5115 265

[3] Y. Afek, G. M. Brown, and M. Merritt. Lazy caching. *ACM Transactions on Programming Languages and Systems*, 15(1):182–205, 1993. DOI: 10.1145/151646.151651 257, 269

[4] J. Alglave, L. Maranget, and M. Tautschnig. Herding cats: Modelling, simulation, testing, and data mining for weak memory. *ACM Transactions on Programming Languages and Systems*, 36(2):7:1–7:74, 2014. DOI: 10.1145/2627752 258

[5] C. J. Banks, M. Elver, R. Hoffmann, S. Sarkar, P. Jackson, and V. Nagarajan. Verification of a lazy cache coherence protocol against a weak memory model. In *Formal Methods in Computer Aided Design (FMCAD)*, pages 60–67, Vienna, Austria, October 2–6, 2017. DOI: 10.23919/fmcad.2017.8102242 266

[6] A. Biere, A. Cimatti, E. M. Clarke, and Y. Zhu. Symbolic model checking without BDDs. In *Tools and Algorithms for Construction and Analysis of Systems, 5th International Conference (TACAS)*, Held as Part of the *Proc. of the European Joint Conferences on the Theory and Practice of Software (ETAPS)*, pages 193–207, Amsterdam, The Netherlands, March 22–28, 1999. DOI: 10.21236/ada360973 265

[7] J. R. Burch, E. M. Clarke, K. L. McMillan, and D. L. Dill. Sequential circuit verification using symbolic model checking. In *Proc. of the 27th ACM/IEEE Design Automation Conference*, pages 46–51, Orlando, FL, June 24–28, 1990. DOI: 10.1109/dac.1990.114827 265

[8] P. Chatterjee, H. Sivaraj, and G. Gopalakrishnan. Shared memory consistency protocol verification against weak memory models: Refinement via model-checking. In *Proc. Computer Aided Verification, 14th International Conference (CAV)*, pages 123–136, Copenhagen, Denmark, July 27–31, 2002. DOI: 10.1007/3-540-45657-0_10 269

[9] K. Chen, S. Malik, and P. Patra. Runtime validation of memory ordering using constraint graph checking. In *14th International Conference on High-Performance Computer Architecture (HPCA-14)*, pages 415–426, Salt Lake City, UT, February 16–20, 2008. DOI: 10.1109/hpca.2008.4658657 268

[10] J. Choi, M. Vijayaraghavan, B. Sherman, A. Chlipala, and Arvind. Kami: A platform for high-level parametric hardware specification and its modular verification. *PACMPL*, 1(ICFP):24:1–24:30, 2017. DOI: 10.1145/3110268 266

[11] C. Chou, P. K. Mannava, and S. Park. A simple method for parameterized verification of cache coherence protocols. In *Proc. Formal Methods in Computer-Aided Design, 5th International Conference (FMCAD)*, pages 382–398, Austin, TX, November 15–17, 2004. DOI: 10.1007/978-3-540-30494-4_27 265

[12] D. L. Dill. The Murphi verification system. In *Proc. Computer Aided Verification, 8th International Conference (CAV)*, pages 390–393, New Brunswick, NJ, July 31–August 3, 1996. 264

[13] M. Elver, C. J. Banks, P. Jackson, and V. Nagarajan. Verc3: A library for explicit state synthesis of concurrent systems. In *Design, Automation and Test in Europe Conference and Exhibition (DATE)*, pages 1381–1386, Dresden, Germany, March 19–23, 2018. DOI: 10.23919/date.2018.8342228 269

[14] M. Elver and V. Nagarajan. Mcversi: A test generation framework for fast memory consistency verification in simulation. In *IEEE International Symposium on High Performance Computer Architecture (HPCA)*, pages 618–630, Barcelona, Spain, March 12–16, 2016. DOI: 10.1109/hpca.2016.7446099 263, 268

[15] S. Hangal, D. Vahia, C. Manovit, J. J. Lu, and S. Narayanan. TSOtool: A program for verifying memory systems using the memory consistency model. In *31st International Symposium on Computer Architecture (ISCA)*, pages 114–123, Munich, Germany, June 19–23, 2004. DOI: 10.1109/isca.2004.1310768 263, 268

[16] M. Herlihy and J. M. Wing. Linearizability: A correctness condition for concurrent objects. *ACM Transactions on Programming Languages and Systems*, 12(3):463–492, 1990. DOI: 10.1145/78969.78972 257, 269

[17] R. Jhala and K. L. McMillan. Microarchitecture verification by compositional model checking. In *Proc. Computer Aided Verification, 13th International Conference (CAV)*, pages 396–410, Paris, France, July 18–22, 2001. DOI: 10.1007/3-540-44585-4_40 265

[18] L. Lamport. Time, clocks, and the ordering of events in a distributed system. *Communications on ACM*, 21(7):558–565, 1978. DOI: 10.1145/359545.359563 264

[19] D. Lee and V. Bertacco. MTraceCheck: Validating non-deterministic behavior of memory consistency models in post-silicon validation. In *Proc. of the 44th Annual International Symposium on Computer Architecture (ISCA)*, pages 201–213, Toronto, ON, Canada, June 24–28, 2017. DOI: 10.1145/3079856.3080235 268

[20] D. Lustig, M. Pellauer, and M. Martonosi. PipeCheck: Specifying and verifying microarchitectural enforcement of memory consistency models. In *47th Annual IEEE/ACM International Symposium on Microarchitecture (MICRO)*, pages 635–646, Cambridge, UK, December 13–17, 2014. DOI: 10.1109/micro.2014.38 260, 266

[21] D. Lustig, G. Sethi, M. Martonosi, and A. Bhattacharjee. COATCheck: Verifying memory ordering at the hardware-OS interface. In *Proc. of the 21st International Conference on Architectural Support for Programming Languages and Operating Systems (ASPLOS)*, pages 233–247, Atlanta, GA, April 2–6, 2016. DOI: 10.1145/2872362.2872399 266

[22] Y. A. Manerkar, D. Lustig, M. Martonosi, and A. Gupta. PipeProof: Automated memory consistency proofs for microarchitectural specifications. In *51st Annual IEEE/ACM International Symposium on Microarchitecture (MICRO)*, pages 788–801, Fukuoka, Japan, October 20–24, 2018. DOI: 10.1109/micro.2018.00069 267

[23] Y. A. Manerkar, D. Lustig, M. Martonosi, and M. Pellauer. RTLCheck: Verifying the memory consistency of RTL designs. In *Proc. of the 50th Annual IEEE/ACM International Symposium on Microarchitecture (MICRO)*, pages 463–476, Cambridge, MA, October 14–18, 2017. DOI: 10.1145/3123939.3124536 269

[24] Y. A. Manerkar, D. Lustig, M. Pellauer, and M. Martonosi. CCICheck: Using μhb graphs to verify the coherence-consistency interface. In *Proc. of the 48th International Symposium on Microarchitecture (MICRO)*, pages 26–37, Waikiki, HI, December 5–9, 2015. DOI: 10.1145/2830772.2830782 261, 266

[25] A. Meixner and D. J. Sorin. Dynamic verification of sequential consistency. In *32st International Symposium on Computer Architecture (ISCA)*, pages 26–37, Madison, WI, June 4–8, 2005. DOI: 10.1109/isca.2005.25 268

[26] A. Meixner and D. J. Sorin. Dynamic verification of memory consistency in cache-coherent multithreaded computer architectures. In *Proc. International Conference on Dependable Systems and Networks (DSN)*, pages 73–82, Philadelphia, PA, June 25–28, 2006. DOI: 10.1109/dsn.2006.29 264

[27] R. Milner. An algebraic definition of simulation between programs. In *Proc. of the 2nd International Joint Conference on Artificial Intelligence*, pages 481–489, London, UK, September 1–3, 1971. 265

[28] N. Oswald, V. Nagarajan, and D. J. Sorin. ProtoGen: Automatically generating directory cache coherence protocols from atomic specifications. In *45th ACM/IEEE Annual International Symposium on Computer Architecture (ISCA)*, pages 247–260, Los Angeles, CA, June 1–6, 2018. DOI: 10.1109/isca.2018.00030 269

[29] S. Park and D. L. Dill. Verification of FLASH cache coherence protocol by aggregation of distributed transactions. In *SPAA*, pages 288–296, 1996. DOI: 10.1145/237502.237573 269

[30] M. Plakal, D. J. Sorin, A. Condon, and M. D. Hill. Lamport clocks: Verifying a directory cache-coherence protocol. In *SPAA*, pages 67–76, 1998. DOI: 10.1145/277651.277672 264

[31] A. Pnueli. The temporal logic of programs. In *18th Annual Symposium on Foundations of Computer Science*, pages 46–57, Providence, Rhode Island, October 31–November 1, 1977. DOI: 10.1109/sfcs.1977.32 254

[32] F. Pong and M. Dubois. Verification techniques for cache coherence protocols. *ACM Computing Surveys*, 29(1):82–126, 1997. DOI: 10.1145/248621.248624 269

[33] S. Qadeer. Verifying sequential consistency on shared-memory multiprocessors by model checking. *IEEE Transactions on Parallel Distributions on Systems*, 14(8):730–741, 2003. DOI: 10.1109/tpds.2003.1225053 269

[34] D. E. Shasha and M. Snir. Efficient and correct execution of parallel programs that share memory. *ACM Transactions on Programming Languages and Systems*, 10(2):282–312, 1988. DOI: 10.1145/42190.42277 269

[35] C. Trippel, Y. A. Manerkar, D. Lustig, M. Pellauer, and M. Martonosi. TriCheck: Memory model verification at the trisection of software, hardware, and ISA. In *Proc. of the 22nd International Conference on Architectural Support for Programming Languages and Operating Systems (ASPLOS)*, pages 119–133, Xi'an, China, April 8–12, 2017. DOI: 10.1145/3037697.3037719 269

[36] A. Udupa, A. Raghavan, J. V. Deshmukh, S. Mador-Haim, M. M. K. Martin, and R. Alur. TRANSIT: Specifying protocols with concolic snippets. In *ACM SIGPLAN Conference on Programming Language Design and Implementation (PLDI)*, pages 287–296, Seattle, WA, June 16–19, 2013. DOI: 10.1145/2491956.2462174 269

[37] M. Zhang, J. D. Bingham, J. Erickson, and D. J. Sorin. PVCoherence: Designing flat coherence protocols for scalable verification. In *20th IEEE International Symposium on High Performance Computer Architecture (HPCA)*, pages 392–403, Orlando, FL, February 15–19, 2014. DOI: 10.1109/mm.2015.48 269

[38] M. Zhang, A. R. Lebeck, and D. J. Sorin. Fractal coherence: Scalably verifiable cache coherence. In *43rd Annual IEEE/ACM International Symposium on Microarchitecture (MI-CRO)*, pages 471–482, Atlanta, GA, December 4–8, 2010. DOI: 10.1109/micro.2010.11 269

Authors' Biographies

VIJAY NAGARAJAN

Vijay Nagarajan (http://homepages.inf.ed.ac.uk/vnagaraj/) is a Reader at the School of Informatics at the University of Edinburgh. He received a Ph.D. in Computer Science from University of California, Riverside. His research interests span computer architecture, compilers, and computer systems with a focus on memory consistency models and cache coherence protocols. He is a recipient of the Intel early career faculty honour award, a PACT best paper award, and an IEEE Top Picks honorable mention. He has served (or is currently serving) on multiple program committees including ISCA, MICRO, and HPCA. He was General Chair of LCTES 2017 and is currently serving as an Associate Editor of *IEEE Computer Architecture Letters (IEEE CAL)*.

DANIEL J. SORIN

Daniel J. Sorin is Professor of Electrical and Computer Engineering and of Computer Science at Duke University. His research interests are in computer architecture, including dependable architectures, verification-aware processor design, and memory system design. He received a Ph.D. and M.S. in electrical and computer engineering from the University of Wisconsin, and he received a BSE in electrical engineering from Duke University. He is the recipient of an NSF Career Award, and he was a Distinguished Visiting Fellow of the Royal Academy of Engineering (UK). He is the Editor-in-Chief of *IEEE Computer Architecture Letters*, and he is a Founder and Chief Architect of Realtime Robotics, Inc. He is the author of a previous Synthesis Lecture, *Fault Tolerant Computer Architecture* (2009).

MARK D. HILL

Mark D. Hill (http://www.cs.wisc.edu/~markhill) is John P. Morgridge Professor and Gene M. Amdahl Professor of Computer Sciences at the University of Wisconsin-Madison, where he also has a courtesy appointment in Electrical and Computer Engineering. His research interests and accomplishments are in parallel-computer system design (e.g., data-race-free memory consistency), memory system design (3C model: compulsory, capacity, and conflict misses), and computer simulation (GEMS and gem5). Hill's work is highly collaborative with over 160 co-authors and especially his long-time colleague David A. Wood. He received the 2019 Eckert-Mauchly Award and 2009 ACM SIGARCH Alan Berenbaum Distinguished

Service Award. Hill is a fellow of IEEE and the ACM. He served as Chair of the Computer Community Consortium from 2018–2020 and as Wisconsin Computer Sciences Department Chair from 2014–2017. Hill has a Ph.D. in Computer Science from the University of California, Berkeley.

DAVID A. WOOD

David A. Wood is Professor Emeritus of Computer Sciences at the University of Wisconsin, Madison, where he was the Sheldon B. Lubar Chair in Computer Sciences, the Amar and Balinder Sohi Professor in Computer Science, and held a courtesy appointment in Electrical and Computer Engineering. Dr. Wood has a Ph.D. in Computer Science (1990) from UC Berkeley. Dr. Wood is an ACM Fellow (2006), IEEE Fellow (2004), UW Vilas Associate (2011), UW Romnes Fellow (1999), and NSF PYI (1991). Dr. Wood served as Chair of ACM SIGARCH, Area Editor (Computer Systems) of *ACM TOMACS*, Associate Editor of *ACM TACO*, Program Committee Chairman of ASPLOS-X (2002), and served on numerous program committees. Dr. Wood has published over 100 technical papers and is an inventor on 19 U.S. patents. Dr. Wood co-led the Wisconsin Wind Tunnel and Wisconsin Multifacet projects with his long-time collaborator Mark D. Hill.

Printed in the United States
by Baker & Taylor Publisher Services